This book

Gas-Solids Handling
in the
Process Industries

CHEMICAL PROCESSING AND ENGINEERING

An International Series of Monographs and Textbooks

EDITORS

Lyle F. Albright

Purdue University
West Lafayette. Indiana

R. N. Maddox

Oklahoma State University
Stillwater. Oklahoma

John J. McKetta

University of Texas
at Austin
Austin. Texas

Gas-Solids Handling in the Process Industries

Edited by

Joseph M. Marchello and Albert Gomezplata

Chemical Engineering Department
University of Maryland
College Park, Maryland

MARCEL DEKKER. INC.   New York and Basel

MARCEL DEKKER, INC.

270 Madison Avenue, New York, New York 10016

LIBRARY OF CONGRESS CATALOG CARD NUMBER: 75-2675
ISBN: 0-8247-6302-5

Current printing (last digit):
10 9 8 7 6 5 4 3 2 1

660·281
GAS

PRINTED IN THE UNITED STATES OF AMERICA

LIST OF CONTRIBUTORS

H. WILLIAM BLAKESLEE, Industrial Emissions Department, Scott Environmental Technology, Inc., Plumsteadville, Pennsylvania

RALPH L. CARR, JR., BIF Chemical Laboratory, BIF, A Unit of General Signal Corp., West Warwick, Rhode Island

J. T. CARSTENSEN, School of Pharmacy, University of Wisconsin, Madison, Wisconsin

ALBERT GOMEZPLATA, Chemical Engineering Department, University of Maryland, College Park, Maryland

G. A. HOHNER, Research and Development, The Quaker Oats Co., Barrington, Illinois

A. M. KUGELMAN, Chemical Engineering Department, University of Maryland, College Park, Maryland

JOSEPH M. MARCHELLO, Division of Mathematical and Physical Sciences and Engineering, University of Maryland, College Park, Maryland

WILLIAM S. O'BRIEN, Department of Thermal and Environmental Engineering, Southern Illinois University at Carbondale, Carbondale, Illinois

C. Y. WEN, Department of Chemical Engineering, West Virginia University, Morgantown, West Virginia

iii

FOREWORD

This Chemical Engineering and Processing Series will be a welcome addition to the literature in that it provides new material on a series of topics important to the chemical engineering aspects of the process industries.

Today we are in great need of carrying new scientific and engineering developments through to the application state. Unfortunately chemical engineering had a lull in attention to new information via university research having to do with the real world of the process industries. It is believed this situation has turned around both in research and teaching and this series of books should assist in the educational effort in this direction. The bridge between science and engineering application is a continuing challenge to chemical engineers and has been taken up by the editors of this series.

One important characteristic of the book or series is the background and ability of the authors and editors, their discernment of what is important and what should be left out, their dedication to the task of communicating the essence of ideas to the reader, and their preparation of material from the viewpoint of the user overrides most other considerations. I wish to commend Marcel Dekker, Inc., for the leadership they have chosen and the editors in turn for prospective lists of authors.

I wish the editors and authors well in their efforts to increase the use of technical knowledge for the benefit of process industries and hence the people of our nation.

DONALD L. KATZ

v

PREFACE

The study of gas-solid systems is an interdisciplinary subject requiring application of physics, chemistry, mathematics, and engineering. Many practical problems involving gas-solid handling have been solved by engineers in the metal, chemical, food, and pharmaceutical industries. This book brings together theory and practice on the handling of gas-solid systems to provide an integrated view and a reference source of the field.

A major part of the process industry is concerned with handling material in powder and granular form. Much of this activity involves processing the solids in contact with gases and vapor and, to an increasing extent, using gas to convey solids. This book is directed toward gas-solid handling as practiced by industry to process and transport materials. The basic aspects of particle and bulk solid mechanics and the design and performance of handling systems are presented with the aim of providing the reader with the background to better handle the many complex problems of the field.

Chapter 1 provides an introduction to the field. Chapter 2 deals with the properties and mechanics of powders and granules, and Chapters 3 and 4 present the principles and practice of pneumatic conveying and of gas-solid processing systems. These chapters provide a basis for the three concluding chapters, which focus on specific applications in the metal, pharmaceutical, and food industries.

The contributing authors possess extensive and multi-leveled experience in gas-solids handling systems. Some have years of experience in industry while others have extensive teaching and research experience. This book draws upon their collective experience and summarizes it for the convenience of future workers.

J. M. Marchello

A. Gomezplata

CONTENTS

Joseph M. Marchello

Chemical Engineering Department
University of Maryland
College Park, Maryland

I. INTRODUCTION

Enormous amounts of solid materials in powder and granular form
are handled and processed using gases. Such activities include size
reduction, conveying, classifying, drying, heating, and chemical
processing. These operations are employed throughout the process
industry for the manufacture of chemicals, metals, foods, and related
products.

Many of the gas-solid operations are highly automated systems
employing the latest methods of control and system design. This book

reviews the basic principles of powder and granular properties and mechanics and relates them to the general operations of transporting and processing and to specific industrial applications.

II. PROCESS INDUSTRIES

The process industry spans an extremely broad and diverse range of products and processes ranging from fuels and construction materials to agricultural products, chemicals, paper, and synthetic fibers [1]. With few exceptions the processes are complex, involving a number of sequential steps in going from raw materials to finished products. These individual steps are generally referred to as unit operations and unit processes.

This generalization of processes into unit operations such as heat transfer, conveying, classification, grinding, and adsorption has permitted their independent study and development [2]. As a result, knowledge gained in one area has often been transferred to another for the mutual benefit of the field.

A. CHEMICAL MANUFACTURE

Many chemical products are in powder or granular form or are transformed through these forms at one or more points in their manufacture [1]. Some examples of products and process operations that involve significant gas-solid handling are:

Portland Cement. Rock, limestone, clay, shale, slag, and mixed to proper proportions and calcined to carry out various chemical reactions. The resulting clinkers are ground in a mill, air classified, and stored for packaging.

Complex Fertilizers. Ground phosphate rock and potassium chloride are treated with ammonia and nitric and phosphoric acids, spherodized with hot gases in a rotary dryer, screened and conveyed to a spray coater, and stored or packaged.

Granule Detergents. Alkyl-benzenes are sulfonated, neutralized,
and dried with hot air in a vertical spray tower, separated in a cyclone, screened, perfumed, and packaged.

Petroleum Cracking. In hydrocarbon cracking operations large
amounts of catalyst in the form of granules or spheres are continuously contacted with hydrocarbon vapors in reactors, transferred to
regenerators, and recycled to the reactor.

B. METAL PROCESSING

A number of gas-solid handling operations are used in ore processing and in primary metal production [3]. Noteworthy among these are
the roasting, agglomeration, and calcination steps of pyrometallurgy.

Metal parts fabrication by powder metallurgy presents a growing
manufacturing area, primarily as a result of the high production rates
that can be achieved through automation. Some of the many gas-solid
interactions involved in powder metal processing are discussed in
Chapter 5.

C. FOOD AND PHARMACEUTICALS

The manufacture of foodstuffs and pharmaceuticals presents a wide
variety of gas-solid operations such as conveying, drying, roasting,
milling, mixing, and blending. These are presented in Chapters 6 and
7. An important added consideration in dealing with materials for
human and animal consumption is the need for microbial control and
the maintenance of sterile and clean conditions.

III. GAS-SOLID HANDLING SYSTEMS

The handling of solids with gases requires a knowledge of the
individual and bulk properties of solids and their mechanical and
chemical behavior. In a broad sense, gas-solid handling systems may

be classified as those dealing primarily with transporting solids and
those involving solids processing steps, Chapters 3 and 4. Size reduction, mechanical conveying, and dust control are important aspects
of solids processing that generally fall outside the scope of this
book. For completeness and to assist the reader they are briefly
reviewed in this section.

A. SIZE REDUCTION

Crushing and grinding, or comminution, involve reducing the size
of solid particles with the attendant creation of new surfaces. The
energy required for size reduction is dependent on amount of reduction,
or new surface formed, as well as on the elastic and plastic properties of the materials [2].

A variety of equipment has been developed for the crushing and
grinding of solids [4]. These include, in decreasing product size:
jaw, roll, and impact crushers; roller, rod, pebble, and ball mills;
disintegrators; and fluid-energy mills. The size reduction of each
type of machine is limited, so that for significant amounts of reduction it may be necessary to carry out the reduction in a series
of stages using several types of machines.

Particle size analysis, classification, and separation are generally carried out by screening, sieving, or for fine particles below
about 60 mesh (250 μm), by cyclone separation [5]. Under crushing
stress, the fracture pattern is influenced by planes of weakness in
the structure of the material, by the presence or absence of fines
in the feed, and by the type of crusher used. The size distinction
of an irregularly shaped particle may be made by reference to its
dimensions, surface area, volume, or weight, or by a comparison of
physical properties that are dependent on these factors. In the finer
sizes, the microscope is used in measuring particle size, or classification may be made by observing differences in the settling rates
of particles in a fluid.

The most widely used method of measurement of the quantity of
material in each size fraction is the weight. The relation between

weight and particle count is influenced by variations in particle
shape and density in the different fractions. In fine fractions
there are often practical problems of obtaining accurate samples and
of measuring particle size. With coarse fractions, the number of
particles present may be small, resulting in sampling deviations.

A graph of weight-size distribution for crushed or ground material
is particularly useful. It can illustrate characteristic features
of the material and aid in comparison of results for experimental
errors or lack of homogeneity in the material.

In the first stages of reduction by crushing, there is a choice
between those machines that reduce by compression between two faces
and those that reduce by impact. There is also some reduction by
attrition, but the particle-size distributions produced by impact and
by attrition show rather similar characteristics, with a wide spread
of sizes and a high proportion of fines.

Discharge from a compression-type crusher is controlled by the
gap of the crusher at the bottom of the crushing chamber. This introduces a size selection into the reduction process and yields a
more closely graded fraction, so that the fines produced by splinter-
ing and attrition constitute a smaller proportion of the whole product.
The wear of the crushing surfaces handling an abrasive material is
likely to be less in a compression-type crusher than in an impact
crusher, but the material must be free flowing and not liable to
pack in the chamber.

The most common jaw crusher design employs a vertical or slightly
inclined fixed jaw and a moving jaw that is opened or closed by a
powered toggle system. The size of a jaw crusher is defined by the
dimensions of the feed opening. Commercial units range in size from
6 to 84 in., with capacities as high as 800 tons/hr for the larger
units. The maximum size reduction obtainable is about 10:1, depending
on the material crushed.

Gyratory crushers use a cone and sleeve design. Eccentric rota-
tory motion crushes material fed into the space between the sleeve
and cone. For a given feed opening a gyratory crusher has about

double the capacity of a jaw crusher, but costs are about twice as
high. The choice of machine usually depends on experience and the
specific material to be handled.

The various impact crusher designs have the common feature of
employing rotating hammers inside a case into which the material to
be crushed is fed. Impact crushers generally give a much higher
reduction ratio than other types of compression crushers. However,
the particle size of the product has a wider spread of size range,
often with a large proportion of fines. Produce size is controlled
by hammer rotor speed, by the apertures between breaker plates, and
by their relationship to the rotor. Two factors that can affect use
of impact crushers are the presence of highly abrasive material and
the presence of plastic or sticky material such as clay. Use of
hardened alloys in the machine helps to reduce abrasion. Similarly,
addition of Water to produce a slurry (wet reduction) often overcomes
sticking and balling problems.

There is a variety of grinding equipment to suit different materials and conditions. Hammer mills, pinned-disk mills, ringroll
mills, tube mills, and ball mills each have their characteristic
features and may be combined with various systems of powder classification. Many difficult materials in the chemical process industries,
such as agglomerates, filter cake, and centrifuged products, can be
ground and dried simultaneously, and cooling arrangements may be used
when grinding heat-sensitive materials. Fluid-bed processes, calciners,
and machines for mixing and blending can be incorporated in a common
system. Specialized plants are designed to meet the requirements for
reducing metallic and nonmetallic ores, phosphates, sulfur, barytes,
and other raw materials for the chemical and plastics industries [4].

Coarse grinding is generally defined as reduction to -10 mesh,
fine grinding to 95% through 200 mesh, very fine to 99.9% through
300 mesh, and superfine to the subsieve size range, or less than 44
μm.

For friable or medium hard material of a free-flowing nature, the
pinned-disk mill can be used for fine grinding with a bagging-off

valve and filter. The mill may be airswept to remove and separate
the fraction required and to reduce overgrinding. A unit, incorporat-
ing grinding and classification, will give a size range from 10 mesh
to 20 μm.

A disintegrator is a type of hammer mill with closed-circuit
grinding. The output is usually fed to an air classifier with fine
product collected in a cyclone. Pulverizers are similar to disintegrators. They are air-swept swing-hammer mills and use a classifier
to reject and return oversize material.

In roller mills grinding is effected by the action running inside
a stationary ring. The rollers move outward due to centrifugal force.
Plows in the grinding chamber project the material between the rollers
and rings. An important feature of roller mills is the ability to
grind in a controlled atmosphere where flammability of the material
is of concern.

Pebble and ball mills use the impact action of metal or ceramic
balls or flint pebbles in a rotating chamber to grind material. Because the balls wear and are replenished, there is a range of sizes
in use. It is advantageous to segregate the sizes so that the large
heavier balls are used at the feed end. This is often done by dividing the drum into sections by the use of ring spacers.

To obtain a closely graded dry product with few fines, it is
advantageous to maintain a continuous circulation of material by
separating the fine particles and returning the oversize. This can
be achieved by passing the discharge through a suitable classifier,
and returning the oversize to the mill feed. Alternatively, a current
of air may be passed through the mill to remove the powder in a closed
circuit, which includes a separator to control the product's particle
size and a cyclone to remove it. Such a system may be kept below
atmospheric pressure to avoid dust leakage. A more granular and
closely graded product can be obtained in this way, and overgrinding
and power consumption are minimized.

Very fine grinding is often carried out using vibratory ball mills
and fluid-energy mills. In vibratory mills the cylinders, horizontal

or vertical, are vibrated so that fine material overflows an adjustable
weir and is passed to a classifier for separation and return of oversize material. The fluid-energy mill produces powders in the subsieve
range, frequently below 5 μm. Grinding is by attrition using compressed gases, air, or steam. Drying and grinding can be done in one
operation. The mill can be lined with a wide variety of materials
to avoid undesirable contamination. It is used for ceramics, plastics,
pigments, pharmaceuticals, insecticides, sterile products, antibiotics,
dyestuffs, heat-sensitive materials, and minerals.

B. TRANSPORTING

Conveying or transporting bulk solids from one point to another
is carried out with mechanical conveyors and elevators and with
pneumatic conveying systems. The choice of system and equipment
depends on the materials to be handled, the path of travel, and other
site considerations. For example, for delivery to remote plant areas
or when dust-tight protection is needed a pneumatic system is frequently selected.

Mechanical conveying and elevating equipment may be classified
in the following categories:

Screw conveyors	Pan conveyors
Belt conveyors	En masse conveyors
Bucket elevators	Vibratory conveyors
Screw elevators	Skip elevators
Drag conveyors	Bucket carriers

Various charts and procedures have been developed by equipment manu-
facturers to help in the selection of the best equipment type for a
given job [6]. The first step is to classify the material according
to density, abrasiveness, corrosiveness, packing tendency, etc. From
such information a preliminary choice of type may be made and more
detailed designs developed.

Limitations to the use of mechanical conveyors are related to
design factors and material properties. Dusting of fine materials

may require equipment enclosures and other controls. Temperature
limitations also often rule out belt and other types of conveyors.
Horizontal screw conveyors are restricted to operating distances of
less than about 200 ft because of torque and shear capacity limitations of materials of construction. On the other hand, belt conveyors
frequently carry materials 1,000 ft or more and may include a reasonable degree of inclination from horizontal.

The design and performance of pneumatic conveyors is presented
in Chapter 3. They are generally used to deliver dry, granular, or
powdered materials via pipelines to remote areas [7]. Since pneumatic systems are enclosed, product contamination, material loss,
and dust emission are reduced or eliminated. Pneumatic conveyors
are often combined with mechanical conveyors and integrated into
processes to increase versatility of handling and control.

C. DUST CONTROL

Few bulk solids handling systems are complete without a dust
control unit. Recent developments in pollution and employee health
regulations emphasize the importance of controlling dust emissions.
Trouble spots include inlet and discharge points and free-fall areas
such as elevator end points, chutes, and storage bins.

Important material characteristics in controlling dust are:
particle size distribution, hygroscopicity, explosiveness, toxicity,
corrosiveness, and electrostatic properties. When specific information on the actual material to be used is not known, it is often
possible to obtain guidance from various literature references dealing
with particle size, corrosion, explosive factors, and possible health
hazards [2,7,8].

There are five general methods for removing dust from air:
(1) settling, (2) centrifugal action, (3) wet scrubbing, (4) filtration, and (5) electrostatic precipitation. Settling techniques
are generally suitable only for large particles or as preliminary
cleaning. Cyclone separators range from large diameter, low efficiency
units to banks of small diameter, high efficiency collectors. Fabric

filters, wet scrubbers, and electrostatic precipitators provide high
efficiency control and are used to control small particle emissions.

The centrifugal force on particles in a spinning gas stream can
be many times greater than gravity. While theoretical analysis of
cyclone performance is incomplete, empirical correlations have been
developed. For example, the collection efficiency has been correlated
in terms of the cut-size D_{pc}, which is the diameter of those particles
collected with 50% efficiency [9]:

$$D_{pc} = \frac{9\mu B}{2\pi N U (\Delta \rho)} \tag{1}$$

where B is the width of the tangential gas inlet to the cyclone, μ is
the gas viscosity, $\Delta \rho$ the difference between the particle and gas
densities, U the inlet velocity, and N is the number of turns that
the gas makes before reversing its flow and leaving the unit. To a
first approximation, N is equal to the cyclone cylinder length divided
by the height of the gas inlet. Collection efficiency for particles
larger than D_{pc} is greater than 50%, while for smaller particles the
efficiency is less.

Fabric filters are used extensively in industrial operations to
recover valuable material, as well as to control air pollution emissions. They are often made in the form of tubular bags or envelopes.
The structure in which the bags hang is known as a baghouse. Small
manually cleaned units handle gas flows of a few hundred cfm, while
large automatically cleaned units have been built to handle over
200,000 cfm. Fabric filters usually provide average collection
efficiencies exceeding 99% at pressure drops ranging from 4 to 6 in.
of water. The amount of filter area required, air-to-cloth ratio,
ranges from 1.5 to 3.0 cfm of gas per square foot of cloth for units
that are cleaned by shaking and 20 to 30 for reverse jet cleaning.

In electrostatic precipitation, the particle-laden gas flows
between a high-voltage electrode wire and a ground collecting surface.
The particles are charged during electrical field breakdown and discharge. The charged particles then drift under the influence of the

field to the collector surface plates or tubes. The collected particles are removed by rapping or vibrating the collector surface, or by
washing the surface with water. Typical gas velocities are 2 to 10
ft/sec; applied voltage ranges from 30 to 100 kV with corona currents
from 0.01 to 1.0 mA/ft of discharge wire.

The efficiency of particle collection, n, in an electrostatic
precipitator is given by

$$n = 1 - \exp(-WA/Q) \tag{2}$$

where W is the particle drift velocity, A the collection area, and Q
the volumetric gas flow rate. Typical values of the drift velocity
encountered in practice range from 0.1 to 0.70 ft/sec, depending on
the properties of the dust and gas. Theoretical values of the drift
velocity are generally two or more times larger than those obtained
in industrial scale precipitators [8].

A variety of equipment has been developed for contacting gases
and liquids to remove particles. Scrubbing generally uses water,
and with reuse, water-solid slurries and mixtures. Low-energy
scrubbers, having 1 to 6 in. of pressure drop, include spray towers,
packed towers, and impingement plate towers. Water requirements may
run 3 to 6 gal/1,000 cu ft of gas and collection efficiencies can
exceed 90% when the particles are largely above 5 to 10 μm. High-
energy scrubbers of the Venturi type impart high velocity to the gas
stream and contact the stream with injected water. Collection efficiencies as high as 99.5% for micrometer size particles may be obtained
for pressure drops in the range of 50 in. of water.

REFERENCES

1. R. N. Shreeve, Chemical Process Industries, 3rd. ed., McGraw-Hill,
   New York, 1967, pp. 11-32.
2. J. H. Perry, ed., Chemical Engineers Handbook, Sections 5,6,7,8,
   and 20, McGraw-Hill, New York, 1963.

3. R. D. Pehlke, Unit Processes of Extractive Metallurgy, American
    Elsevier, New York, 1973, pp. 7-23.
4. A. Ratcliffe, Chem. Eng., 79, No. 15, 62-75 (1972).
5. C. W. Mathews, Chem. Eng., 78, No. 4, 99-104 (1971).
6. M. S. Buffington, Chem. Eng., 76, No. 22, 33-49 (1969).
7. M. N. Kraus, Chem. Eng., 76, No. 22, 59-65 (1969).
8. J. M. Marchello, Control of Air Pollution Sources, Marcel Dekker,
    New York, 1974.
9. J. A. Danielson, ed., "Air Pollution Engineering Manual," Dept.
    of Health, Education and Welfare, U.S.P.H.S. Pub. No. 999-AP-40,
    1967, pp. 25-99.

Chapter 2

POWDER AND GRANULE PROPERTIES AND MECHANICS

Ralph L. Carr, Jr.

BIF Chemical Laboratory
BIF, A Unit of General Signal Corp.
West Warwick, Rhode Island

I. INTRODUCTION [10-14]

A knowledge of the properties and characteristics of individual particles and also of their bulked mass is an essential tool for the process engineer. In the chapters that follow, this knowledge of particles will enable an engineer to more readily understand and follow the unit processes that are spelled out.

Table 1 describes, in compact form, twelve unit operations. The individual particle properties that are important and/or critical to each unit operation have been checked off. Notice that size surface area has been checked off for all unit operations and is also of critical import in every case. Hardness, hygroscopicity, and cohesion follow closely in significance.

TABLE 1

Properties Critical or Important to Each Unit Operation[a]

Unit operation	Size-surface area	Surface activity	Hardness	Melting point	Hygroscopicity, moisture	Angle of repose	Angle of slide	Cohesion	Flow-ability
Size reduction	√C	√	√C	√	√				
Conveying	√C	√	√C	√	√	√C	√C	√	√C
Pneumatic conveying	√C	√	√		√C			√C	√C
fluidization	√C	√	√		√C			√C	√C
Feeders	√C	√C	√C	√	√C			√C	√C
Mixing blending	√C	√	√	√	√	√		√C	
Agglomeration granulation	√C	√	√C	√	√			√	√C
Sintering	√C	√	√	√	√			√	
Drying	√C	√	√	√	√			√	
Additives	√C	√	√	√	√				√
Coating	√C	√	√	√	√			√	
Storage	√C	√C	√C	√	√C	√C	√	√	√
Classification	√C	√	√		√			√	
Particle sizing	√C	√	√		√			√	

[a]C = critical, √ = important property.

The interrelated properties of surface area and size are especially important to the overall field of gas-solids handling. In each case a particular size particle with its resultant surface area will give a certain result. We will first deal with the properties and characteristics of individual particles. Let us define our particle. It is an individual entity, apparently stable because of at least moderate cohesive forces, with its boundaries established by the optical means used to view it. In gas-solids handling, the pertinent range of particle size is from the less familiar colloidal realm to the more familiar granule-crystal level.

II.   INDIVIDUAL PARTICLE PROPERTIES

A.   PARTICLE SIZE, SURFACE AREA, AND SURFACE FORCES [1-9]

Small-sized particles tend to have a relatively greater surface area than larger-sized particles. This relative surface area is the most important property of an individual particle. The size at which particle cohesive forces predominate over the forces of particle free flow or gravity flow depends mainly on the density and molecular weight of the particle. Cohesive particles will constitute the powdered fraction, and the free-flowing units the granule percent. To classify a material into its granular and powdered particles, a screen or sieve analysis is made. For average weight materials (25 to 40 lb/cu ft) of average molecular weight (generally inorganic), the minus 200 mesh will be the cohesive, powdered particles, and the plus 200 mesh will constitute the free-flowing granules. For lighter weight materials, the powder will be the minus 100 mesh and for heavier materials, the minus 325 mesh will be the powder fraction and vice versa. Table 2 illustrates the interrelationship between grade or size of particle and its bulk density and flow properties for a powdered, granular-shaped material.

A study of Table 2 will show the point at which a given property, such as mesh size, will cause a material to have either cohesive,

TABLE 2

Sample of a Powdered Granular Shale with Its Properties
and the Properties of Four Screened Samples of Same[a]

Shale grade or size	Mesh size	Bulk density (lb/cu ft)	Cohesion (%)	Flow	Flood	Compressibility (%)
1. Powdered granules	+ 10 to -200	83 to 105	0	Passable	No	21
2. Granules	+ 10 to + 40	80 to 98	0	Fair	No	18.5
3. Granules	- 10 to + 60	70 to 85	0	Good	No	17.5
4. Granules and powder	- 60 to -200	58 to 80	5	Passable to poor	Yes	27
5. Fluid powder	-200 to -325	49 to 74	12	Poor	Very	33

[a]Powdered, granular shale sample is number 1. Numbers 2 to 5 are screened samples of same.

non-free-flowing particles or noncohesive, free-flowing granules. For
the shale material, this change occurs at 325 mesh. Also, the powder
fraction has a lesser bulk density and a greater compressibility.

Table 3 effectively shows the relationship of porosity, surface
area, and particle size to five unit operations. The materials with
no free flow are the nonporous particles with sizes less than 1 μm:
the small diameter results in relatively large surface areas. Also,
these fine, cohesive, agglomerated masses are more difficult to fluidize,
blend, and hold in storage without massing. More optimum handling is
assured with larger sized particles. Porosity also leads to less
cohesive units.

As surface area for a particle increases, so also will its surface
energy and surface activity. At some point of particle size reduction,
unsatisfied forces already present or forces induced because of broken
bonds or strained lattices will come into play, and free-flowing
granules will become less free-flowing or non-free-flowing powdered
particles. Many of these forces are the familiar van der Waals or
London forces. Such forces will cause cohesion and adhesion of particles. Electrostatic and magnetic forces between solid particles are
important in some instances. The electrostatic forces have more
relevance with dielectric materials. Magnetic forces, of course, are
of significance with magnetic substances.

The following steps can be taken to reduce the electrical activity:
1. Thoroughly ground the material
2. Change the method of conveying
3. Control humidity - limited to materials not affected by
    moisture
4. Change equipment material - equalize charges, etc.
5. Use antistatic agents - subject to surface loss
6. Use additives - must be compatible with the material
7. Reduce surface impurities - if practical
8. Change flow rate - high rates lead to ineffective distribu-
    tion of charges

The free surface energy of a particle is a very important property
to observe. It is the work involved in forming a unit area of the

TABLE 3

Effect of Size-Surface Area and Porosity
on the Listed Unit Operations

Material	Porous[a]	Range of Particle size (μm)	Surface area (m²/g)	Flowable	Floodable	Conveying	Fluidizable	Mixing-blending	Storage
Calcium carbonate precipitate	Non	0.6-0.8	2- 5	Very poor	No	Very cohesive, packs	Some difficulty	Some difficulty	Very compressible
Zinc oxide	Non	0.6-10	10- 35	Very poor	No	" "	" "	" "	"
Iron oxide	Non	0.06-0.2	10-100	Very poor	No	Cohesive packs	" "	" "	"
Titanium dioxide	Non	0.2-0.4	10- 15	Very poor	No	Very cohesive, packs	" "	" "	"
Carbon black	Non	0.003-0.05	60-1,000	Very poor	No	Very cohesive dust	Fluidizable	Yes	"
Carbon, activated	Yes	5 - 40	600-750	Poor	Very	Aerates,[e] dusty	Readily	Aerates	Aerates
Fullers earth	Yes	0.06-0.08	150	Passable	Very	" "	" "	" "	"
Fullers earth	Yes	0.3-0.4	125	Passable	Yes	" "	" "	" "	"
Santocel fine silica	Yes	3 - 5	500-550	Very poor	Can be fluidized	Very dusty too light	Very dusty	Very dusty	Bulky
Copper sulfate, hydrated	Non	100-300	0.1-0.2	Fair	No	Efflorescent	No	No	Efflorescent
Sand, coarse silica	Non	60-100	0.2-0.5	Excellent	No	Abrasive F.F.[c]	No - will air convey	"	No problem
Alumina, activated	(Yes)[b]	400-1,000	100-2,000	Good	No	" "	" "	" "	"

[a] Porous particles have both an external and an internal surface area.

[b] Partial.

[c] FF = freeflowing.

particle's surface. Units are W/A in ergs/cm^2. With liquids, surface
energy and surface tension are numerically similar. The determination
of the surface energy of solid particles, however, presents a problem.
Hard, high melting-point particles have surface energies of from 500
to 5,000 ergs/cm^2. Softer, inorganic solid units have surface energies
from 100 to 500 ergs/cm^2, and soft organic materials have less than
100 ergs/cm^2. For example, the diamond has a surface energy of over
5,500 ergs/cm^2, whereas salt has only 155 ergs/cm^2. The softer the
particulate unit, with its relatively lower surface energy, the greater
effect there will be with cohesiveness with subdivision.

Another very important surface force is the effective electrical
activity [5,15] that can be present in the relatively large surface
area of smaller-sized particles. Here, electrostatic activity can be
produced on particles by the sudden separation of their surfaces,
especially with dissimilar materials. As a solid loses electrons by
such action, its surfaces acquire an excess positive charge. Attraction
to other charged surfaces is now possible. Especially nonconductive
units, such as plastics, can acquire charged surfaces. Unit operations
of conveying, classifying, mixing, or feeding through a hopper can
present problems in electrostatic activity.

Electrostatic charge properties of particles are important for
flowability, conveying, fluidizing, mixing, blending, and agglomeration.

Another kind of surface activity can occur with larger-sized
particles as well as small particles. This activity occurs in materials
in an unstable form that revert to a more stable form through the
process of hydration, oxidation, or sorption. This activity on particle
surfaces can cause problems in particle handling or movement. Massing
can occur with such materials.

In summation: particle size, surface area, and activity are the
most important particulate measurements, as basic properties. The
size-surface area ratio and the chemical reactions that occur on the
surface are of paramount importance in each unit operation. See Table
1 and note that the size-surface area measurement is checked off for
every unit operation.

B. DENSITY

The density, or the specific gravity, of a material is readily available, or it can be readily determined. However, this basic property cannot directly indicate the handling properties of a substance, although specific gravity may be used to indicate the possible bulk density or cohesive possibilities. Therefore, it is the apparent density of an agglomeration of particles that is referred to here. A granule is a dense, free-flowing, noncohesive entity; while a powdered particle is a much less dense, more cohesive, non-free-flowing entity. Denser agglomerated units tend to stay together and are more free-flowing. They are harder to fluidize also.

If the actual density of each material is checked, the lower-density particles are found to be bulkier than the higher-density substances.

C. HARDNESS AND DEFORMABILITY

How hard a particle is also indicates how deformable it is. Harder particles are less deformable. The extent to which there is cohesion or massing together of particles into a nonflowing agglomeration depends on the deformability of the individual particles. Softer, more deformable particles mass together to a greater extent, and harder, less deformable units to a lesser extent.

Mohs mineral scale of hardness is universally used to relate the deformability of materials. The scale runs from 1 to 10, the softest material having a designation of 1, with talc as the representative mineral. The diamond, being the hardest mineral, is given the highest number, 10. However, this scale is not linear; Table 4 shows, in compact form, the Mohs hardness scale values along with true linear values. Notice especially the jump in actual hardness that occurs between 9.5 and 10. Listed are the more prevalent industrial materials along with the 10 minerals used in the Mohs table. Thereby, such materials can be related to a hardness value.

In general harder particles will be more difficult to (1) reduce in size, (2) blend, (3) agglomerate and granulate. They will be less

TABLE 4

Mohs Hardness Scale and Relative Linear Scale[a]

Materials[b]	Mohs scale	Relative linear scale	Materials[c]	Mohs scale	Relative linear scale
b Wax (0°)	0.02	0.006	Barite	3.3	0.21
Graphite	0.05-1.0	0.02	Barium carbonate	3.5	0.24
a Talc	1.0	0.03	Copper oxide	3.5	0.24
Ammonium nitrate	1.0	0.03	Sodium fluoride	3.5	0.24
c Diat. earth	1.0 -6.0	0.03-0.8	Dolomite	3.5-6.0	0.24-0.8
Asphalt	1.5	0.04	a Fluorite-fluorspar	4.0	0.25
Ice (0°C)	1.5	0.04	Calcium chloride	4.0	0.25
d Lead	1.5	0.04	Magnesium carbonate	4.0-	0.25
Bentonite clay	1.5 -2.0	0.04-0.05	b Platinum	4.0-4.5	0.25-0.4
a Gypsum	2.0	0.05	d Titanium and alloys	4.0-6.0	0.25-0.8
b Fingernail	2.0	0.05	d Iron	4.5	0.4
Aluminum sulfate	2.0	0.05	Zinc oxide	4.5	0.4
Ammonium chloride	2.0	0.05	d Hast. C	4.5-5.0	0.4 -0.6
Cocoa	2.0	0.05	d Glass	4.5-6.5	0.4 -0.1
Flours	2.0	0.05	d Alloy steels	4.5-7.0	0.4 -1.2
Foods in general	2.0 -1.0	0.05-0.03	b Palladium	4.8	0.5

Material		
Iron sulfates	2.0	0.05
Organic crystals	2.0	0.05
Plaster of Paris	2.0	0.05
Potassium chloride	2.0	0.05
Selenium	2.0	0.05
Soda ash	2.0	0.05
Sodium aluminate	2.0	0.05
Sodium nitrate	3.0	0.05
Sodium silicofluoride	2.0	0.05
Starches	2.0	0.05
Sugars	2.0	0.05
Sulfur	2.0	0.05
Tin	2.0	0.05
Zinc	2.0	0.05
Epsom salts	2.0 -2.5	0.05-0.13
Hydrated lime	2.0 -3.0	0.05-0.20
Aluminum and alloys	2.0 -3.0	0.05-0.20
Clays	2.0 -3.0	0.05-0.20
Limestone	2.0 -4.0	0.05-0.25
Quicklime	2.0 -4.0	0.05-0.25
Copper coin	2.5	0.13

Material		
a Apatite	5.0	0.6
Asbestos	5.0	0.6
Calcium phosphates	5.0	0.6
Carbon black	5.0	0.6
Magnesium oxide	5.0-6.5	0.6 -1.0
Kyanite	5.0-7.0	0.6 -1.3
b Knife blade	5.2	0.65
b Window glass	5.2	0.65
c Chromite	5.5	0.7
Mn and Mn dioxide	5-5.5	0.6-0.7
d Steel	5.0-8.5	0.6 -2.3
a Orthoclase-feldspar	6.0	0.8
Pumice	6.0	0.8
b Iridium	6.0-6.5	0.8 -1.0
Molybdenum	6.0-7.0	0.8 -1.3
b File	6.3	0.9
Iron oxides	6.0-7.0	0.8 -1.3
d Hard rubber	6.0-7.0	0.8 -1.3
Aluminum silicates	6.0-7.5	0.8 -1.5
d Tantalum	6.0-7.0	0.8 -1.2
Pyrites	6.5	1.0

TABLE 4 (cont.)

Materials[b]	Mohs scale	Relative linear scale	Materials[c]	Mohs scale	Relative linear scale
Anthracite	2.5	0.13	Titanium dioxide	6.5	1.0
Ammonium sulfate	2.5	0.13	Tungsten	6.5-7.5	1.0 -1.5
Borax	2.5	0.13	a Quartz-silica	7.0	1.25
Copper sulfate	2.5	0.13	Sand	7.0	1.25
Litharge	2.5	0.13	Silicon	7.0	1.25
Magnesium and alloys	2.5	0.13	Beryllia	7.0+	1.25+
Phenolics	2.5	0.13	Zirconia	7.0+	1.25+
Polyester	2.5	0.13	Emery	7.0-9.0	1.3 -2.0+
Potassium permanganate	2.5	0.13	Garnet	7.0-8+	1.3 -2.0
Salt	2.5	0.13	a Topaz	8.0	2.0
Salt cake	2.5	0.13	d Alumina ceramic	8.0	2.0
Silver	2.5	0.13	Beryllium carbide	8.7	2.3

Material			Material		
Sodium bicarbonate	2.5	0.13	a Corundum	9.0	2.5
Copper	2.5 -3.0	0.13-0.20	Chromium	9.0	2.5
Gold	2.5 -3.0	0.13-0.20	Titanium boride	9.0+	2.5+
Sodium phosphates	2.5 -3.0	0.13-0.30	Alundum	9.0+	2.5+
Nickel and alloys	2.5 -5.0	0.13-0.60	Tungsten carbide	9.2	2.7
Calcite	3.0	0.2	Alumina	9.25	2.9
Bauxite	3.0	0.2	Tantalum carbide	9.3	3.0
Plastics, generally	3.0	0.2	Titanium carbide	9.4	3.5
Mica	3.0 -2.5	0.2 -0.13	Silicon carbide	9.4	3.5
Monel	3.0 -5.0	0.2 -0.60	Boron carbide	9.5	4.0
Beryllium	3.0 -7.8	0.2 -1.70	Cubic boron nitrate	10.0	10.0
Brass	3.0 -4.0	0.2 -0.25	a Diamond	10.0	10.0

[a] Adapted from Chemical Engineering, by special permission, Oct. 13, 1969. McGraw-Hill, Inc., N. Y., N. Y.

[b] a = mineral standard; b = Used to relate hardness of other materials.

[c] c = material can have more abrasive impurities; d = material of construction as well.

of a problem in (1) conveying and feeding (except for abrasion), and
in (2) storage and classification. The relative hardness value of a
material can be used to predict: (1) Its storage; materials with a
hardness of less than 4 are more likely to mass. (2) Arching in
hopper; materials with a hardness of less than 4 are more likely to
mass. (3) Flowability; the harder a particle, the more flowable.
(4) Fluidization; the softer the particle, the more agglomerable and
harder to fluidize. (5) Abrasiveness; the harder, the more abrasive,
hardness values >5 are critical [16]. (6) Size reduction; the
harder, the more energy needed to reduce. (7) Handling; in general,
the harder the particle, the easier to handle. (8) Blending; the
harder, the more difficult to blend.

The following materials have hard to very hard, abrasive particles.
The larger this particle, the more critical its abrasiveness or hardness
will be:

Aggregates

Alumina or aluminum oxide or
corundum

Calcined materials

Carbides

Cement

Ceramics

Chromite

Clinkers, cinders

Glass, cullet, or glass batch

Gravel

Metal powders such as iron,
molybdenum, beryllium, and
tungsten

Ores in general

Sand, silica, or silicon dioxide

Shale

Slag

Titanium and compounds

Tungsten carbide

The following are very soft materials which can have problems in
flow, fluidization, mixing, drying, and storage:

Bicarbonate of soda

Cocoa

2,4D acid

Flour

Graphite

Organic chemicals in general

Potassium chloride

Sulfur

Sodium silicofluorides

Talc

Many organic pharmaceutical chemicals

Many organic food substances

The hardness of a particle, which also includes its related surface energy, is, following surface area, its most important property.

D. RUGOSITY AND SHAPE

The ratio of the actual measured, external surface area of a particle to its hypothetical area, as if it were a perfect sphere, is a measure of its rugosity. The reciprocal is sphericity. The greater this difference, the more critical the following properties can be, excluding other factors:

1. Flowability; archability, massing (more points of contact and more cohesion with softer materials)
2. Mixing; blending, hardness
3. Conveying; hardness
4. Granulation; hardness
5. Storage; hardness

Spherical or roundish-shaped particles and squarish shapes also will tend to be more flowable and have less of a tendency to mass. While particles of irregular or random shape, having many intraparticle contacts, tend to be less flowable and have a greater massing tendency. Acicular, flat or mica-like, fibrous, angular, and columnar-shaped units also tend to be less flowable. Shape has not been used directly in evaluating the flow of a particle. The effect of shape is indirectly indicated by other measured solids properties, such as angles of repose and friction, cohesion and shear force, and compressibility. The hardness of the particle would be the most important other single factor in each case.

E. HYGROSCOPICITY [2]

This is the general, all inclusive term which denotes the tendency for particles to attract moisture from ambient surroundings. Following are qualities that define the more specific hygroscopic nature of a material.

1. Deliquescence
    Very soluble salts such as nitrates of sodium and ammonium;
chlorides of aluminum, calcium, and zinc; hydroxides of sodium and
potassium; sodium dichromate; and sodium cyanide all show deliquescence as a critical property. Their crystals tend to pick up enough
moisture from the ambient atmosphere in which to dissolve.

2. "Cake-Up"
    The materials in Table 5 do not pick up enough moisture to dis-
solve, but they "cake up" into various degrees of hardness, with the
sorption of moisture, depending on the relative humidity of the atmosphere.

3. Critical Humidity
    Many crystalline salts or chemicals have a critical humidity
above which they will absorb moisture and cake up or deliquesce. The
ambient temperature is also a factor. The very deliquescent alkali,
sodium hydroxide, has a critical humidity of 59% at 20°C. Some others
are listed in Table 6.

TABLE 5
Materials That Tend to "Cake-Up"

Relative humidity <80%	Relative humidity >80%
Ammonium chloride	Aluminum $14H_2O$
Ammonium sulfate	Bicarbonate of soda
Potash, impure	Ferric sulfate
Soda ash	Milk powder
Sodium aluminate	Potassium permanganate
Sodium chromate	Salt, impure
Sodium monophosphate, anhydrous	Saltcake
Sodium diphosphate, anhydrous	Urea

TABLE 6

Critical Humidity of Salts and Chemicals

Chemical	Temp. (°C)	Relative humidity (%)
$CaCl_2 \cdot 6H_2O$	20	32
KNO_3	20	45
$Na_2Cr_2O_7 \cdot 2H_2O$	20	52
$CaCl_2$	20	58
$AlCl_3$	20	58
$NaNO_3$	20	66
$NaHCO$	25	68
$NaHSO_3$	25	70
$NaClO_3$	20	70
NH_4NO_3	10	75
$NaCl$	20	76
NH_4SO_4	25	81.1
NH_4SO_4	20	81
NH_4SO_4	10	80
Na_2CO_3	25	85
Na_2SO_4	25	85
KNO_3	30	9
$Al_2(SO_4)_3 \cdot 14H_2O$	25	95
NaF	100	96
$CuSO_4 \cdot 5H_2O$	25	97

4. Adsorption, Sorption, or Hygroscopic Moisture or Moisture Regain

The equilibrium moisture content is a way to describe what happens to material particles that do not deliquesce or cake, but do have a change in their properties of flow, etc., with the sorption of moisture. Flour, starch, asbestos, gypsum, hydrated lime, and superphosphate are among such materials.

5. Efflorescence

Some hydrated "crystalline" chemicals instead of picking up surrounding moisture will lose their water of crystallization or hydration when they are exposed to relatively dry air. Temperature is also a factor. The crystals actually will crumble to a whitish, mealy powder. The following hydrated chemicals will effloresce at a certain low relative humidity and a certain temperature:

Borax	. $10H_2O$	Sodium phosphate (tri)	. $12H_2O$
Epsom salts	. $7H_2O$	Sodium sulfate	. $10H_2O$
Ferrous sulfate	. $7H_2O$	Sodium thiosulfate	. $5H_2O$
Nickel sulfate	. $7H_2O$	Zinc sulfate	. $7H_2O$
Sodium carbonate	. $10H_2O$		

6. Efflorescence and Deliquescence

These hydrated salts can either effloresce or deliquesce according to the ambient humidity and temperature:

Citric acid	. $1H_2O$	Sodium phosphate (mono).	$12H_2O$
Copper chloride	. $2H_2O$	Sodium phosphate (tri)	. $12H_2O$
Copper sulfate	. $5H_2O$	Sodium sulfide	. $9H_2O$
Ferrous sulfate	. $7H_2O$	Sodium sulfate	. $10H_2O$
Sodium carbonate	. $12H_2O$	Sodium thiosulfate	. $5H_2O$

They are efflorescent at a lower and deliquescent at a higher relative humidity.

7. Moisture Effects

Some particles, if dried completely, exhibit increased cohesive forces. This is because interfering films of air and moisture are removed, allowing particles to come closer together. Some clays can have increasing, then decreasing, and then increasing friction (or cohesion) between their particles as the moisture content increases.

Excessive moisture causes a problem in the fluidization of particles that are hygroscopic to some degree. Drier particles are much more readily fluidized. The unit operations of mixing-blending,

agglomeration, granulation, drying, adding materials, and classification, as well as the operations already discussed, must take into
consideration this all important property of hygroscopicity.

F. MELTING POINT

A knowledge of the melting point of the particles of a substance
is also very important in many unit operations. Many organic chemicals
and some hydrated inorganic chemicals have a relatively low melting
point. The low mp substances have "active surfaces" at only about
40% of this melting point.

G. DUSTS EXPLOSIVE WITH AIR [15,17-19]

The explosive tendency of a dust of a certain material in combination with air is an important property which should be known prior to
handling, especially in conveying operations and in handling foodstuffs.
In these cases explosion-proof electrical equipment is a necessity.
Tables 7-9 list compilations of a large number of the materials that
can create an explosive dust condition with air. It is the properties
of surface area and/or combustibility of the individual particles
that are critical here. Very fine combustible particles with expan-
sive surface areas are involved especially. Particles larger than
100 mesh are not considered hazardous, and particles larger than 40
mesh are not considered explosive in combination with air unless they
are chemically unstable. The most critical size levels would be from
about 5 to 70μm. Also there is a minimum and maximum range of dust
concentration in air that is critical for each dusty material. There
are four procedures available to the engineer whereby he can prevent
a possible catastrophe:

1. Keep dust accumulation below the minimum or, mainly, prevent
the formation of a dust cloud above the minimum value.
2. Eliminate all possible sites of ignition.
3. Lower the O_2 content of the air.
4. Add an inert material, making the dust cloud inert.

TABLE 7

Explosive Dusts with Air[a]

Class	Material	*Total[b] No of Explosions	Index		
			Ignition sensitivity max. (or ave)	Explosion severity max. (or ave)	Explosibility max. (or ave)
Agricultural	Cocoa		1.1	1.3	1.4
	Coffee	12	0.4	0.1	<0.1
	Grain	205	2.8	3.3	9.2
	Milk, skim		1.6	0.9	1.4
	Starches	55	10.0	7.0	50.0
	Sugar	28	5.5	2.4	13.2
	Wheat		1.3	1.9	2.5
	Wheat flour	88	2.1	1.8	3.8
Carbonaceous	Act. carbon		<<0.1	0.9	<<0.1
	Asphalt		2.8	2.2	6.2
	Carbon black		<<0.1	-	<<0.1
	Coal, high vol.	56	1.5	1.0	2.0
	Coal, low vol.		<<0.1	-	<<0.1
	Coke		<<0.1	-	<<0.1

	Material				
	Gilsonite	6.9		1.5	>10.0
	Graphite	<<0.1		-	<<0.1
	Pitch, coal tar	2.0	5	1.0	2.5
Metals	Aluminum	1.0		5.0	>10.0
	Chromium	0.1		-	0.1
	Copper	<0.1		1.0	<<0.1
	Iron	0.2		0.1	<0.1
	Magnesium	0.8		1.0	>10.0
Ores (81)	Manganese	0.15		0.15	0.1
Minerals	Pyrite	<0.1		-	<0.1
	Sulfur	2.0+	35	-	1.0+
	Tin	0.2		3.0	0.1
	Titanium	5.0		3.0	>10.0
	Zinc	<0.1		<0.1	<0.1
	Acetal	6.5		1.9	>10.0
	Acrylamide	4.1		0.6	2.5
	Acrylonitrile	8.1		2.3	>10.0
	Cellulose	1.0		2.8	2.8
	Cellulose acetate	3.0		2.0	>10.0
	Epoxy	6.0		2.0	1.9->10.0
	Flurocarbon	<<0.1		None	<<0.1

TABLE 7 (cont.)

Class	Material	Total[b] No. of explosions	Index Ignition sensitivity max. (or ave)	Explosion severity max. (or ave)	Explosibility max. (or ave)
	Gum arabic		0.7	1.4	1.0
Plastics(35)	Lucite-Plexiglas		10.0	2.0	6.0->10.0
Polymers	Nylon		4.0	2.0	4.0->10.0
Organics	Phenolic		5.0	2.0	<0.1->10.0
Resins(5)	Polyethylene		8.0	1.5	3.5->10.0
	Polypropylene		3.0	1.0	<0.1- 10.0
Rubbers	Polyurethane		8.0	1.6	>10.0
Woods(146)	Polyester		3.0	2.0	4.9->10.0
	PVC or vinyl		<<0.1-5.0	<0.1-1.7	<0.1->10.0
	Rosin		34.0	4.0	-
	Rubber	15	6.0	1.2	<0.1->10.0
	Rubber, chlorin.		<<0.1	-	<<0.1
	Styrene		4.0	1.5	0.9->10.0
	Urea		<<0.1	-	<<0.1
	Urea formald.		0.6	1.7	1.0
	Wood		3.0	3.0	7.0->10.0

[a]From Bureau of Mines Bulletin and other sources.

[b]From 1950 to 1959 - number in parentheses = total explosions.

TABLE 8

Explosion Hazards Indexes

Explosion hazard adjective rating	Ignition sensitivity index	Explosion severity index	Explosibility index
Weak	<0.2	<0.5	<0.1
Moderate	0.2-1.0	0.5-1.0	0.1- 1.0
Strong	1.0-5.0	1.0-2.0	1.0-10.0
Severe	>5.0	>2.0	>10.0

H. OTHER PROPERTIES

Other properties of solids, such as corrosivity, flammability, odor, color, specific heat, conductivity, and magnetism will not be discussed here. Corrosivity and flammability are important qualities to consider when a material is to be handled. Especially with corrosive materials that are also hygroscopic, the correct choice of materials of construction is important.

I. SUMMATION

For each operation there are ideal properties that individual particles of a material can have which will make for successful application. As summarized in Table 1, there are properties of individual particles that are important and/or critical to each unit operation. An engineer must familiarize himself with the properties of the materials to be used in his process or unit operation, and if the design of the unit being used is known, the optimum particle properties should be obvious.

TABLE 9

Other Explosive Dusts Listed[a]

Plastics (35)	Agricultural
Polymers	Malt (23)
Organics	Bark (15)
Resins (5)	Spices
Rubbers	Cereals (54)
Woods (146)	Tobacco
Shellac	Cork (40)
Casein	Fertilizers (28)
Pthalic anhydride	
Sodium resinates	
Others	Carbonaceous
Napalm	Lignite
Soaps	Peat
Stearates	Charcoal
Blood flour	
Waxes	Metals (81)
Paper (10)	Ores
Cotton	Minerals
Misc. (52)	Tantalum
Phono record (6)	Silicon
Seed (11)	Vanadium
	Zirconium
	Ferro Silicon
	Calcium silicide

[a]From 1950 to 1959 - number in parentheses = total explosions.

III. BULK SOLIDS OF PARTICLES "EN MASSE" PROPERTIES*

A. BULK DENSITY AND POROSITY

The bulk density of a material is a readily obtained value. Its use, however, in determining, say, a flow evaluation is not feasible. There is no direct relationship between a material's bulk density and its flow characteristics. A heavy substance with a bulk density of 80 lb/cu ft can have the same nonflow characteristics or values as a light material with a bulk density of 20 lb/cu ft. For gas-solid unit operation, the bulk density of a material gives valuable information on:

1. Capacity needed; use aerated or loose bulk density.

2. Capacity rates; use working bulk density.

3. Bulkiness of a material; more bulky, less flowable and vice versa. Less bulking, less storage space needed.

4. Compressive strength possible in a mass; more compressible, more strength. Compressibility is determined through the use of bulk density values. This will be covered later in the chapter.

5. Particle size of a given material at which it is powdered or granular; bulk density is an indication.

6. Whether the material can be fluidized; bulk density is one indication.

7. Possible difficulty in pneumatic conveyance; greater bulk density, more energy needed (for lift).

8. The borderline between floodable and flowable masses. Bulk density is an indication. Powder fraction may be floodable, and granular units may not.

9. How abrasive a mass will be. The heavier the mass, the more accelerated the abrasiveness.

----

*For all lists of materials that are given for examples of a particle property, there can be other grades of a given substance which are heavier, lighter, better flowing, etc.

10. Energy needed to reduce size. The heavier or denser, the
    more energy needed.
11. Energy needed to mix or blend. The heavier the mass, the
    more energy needed.
12. Possible heat exchange. The less bulky the mass, the smaller
    the surface area for heat exchange and vice versa.
13. Possible difficulty in classification. The greater the mass,
    the denser the particles.

Specific gravity is a relative number which relates the mass of
a volume of one particle to the mass of the same volume of water at
a specified temperature. Bulk density is the weight in lb/cu ft
(g/cm^3) of a mass of particles in a specified volume in relation to
the weight of the same mass of water, at a specified temperature.
The bulk density is actually the weight of a mass of particles in a
specified volume. The definition of porosity is:

$$\% \text{ porosity} = 100 - \frac{\text{Average B.D. of material}}{\text{Sp.G. of material x 62.3 (wt. of a cu ft of water)}} \quad (1)$$

or

$$\text{voidage} = 1 - \frac{\text{B.D. (apparent density of the material)}}{\text{Sp.G. (true density of same material)}} \quad (2)$$

Thus, as shown by Eqs. (1) and (2), the porosity of a given
material is the deviation of the bulk weight, or bulk density, (of a
mass of its particles) from the true density (or the relative weight
of one of its particles). The greater this deviation, the greater the
porosity will be within the mass. Also, the greater will be the mass
flow properties of this material. Very cohesive, nonflow materials,
such as zinc oxide, barium carbonate, calcium hydroxide, iron oxide,
and titanium dioxide, have a porosity of over 80%; while a grade of
sand with free-flowing granules has a porosity of only 45%. Generally,
free-flowing granular materials have a porosity of between 35 and 50%.

The determination of the overall B.D. of a material includes the following bulk densities:

1. Loose or aerated B.D.
2. Packed B.D.
3. Working B.D.
4. Average B.D.
5. Fluid B.D. (indicated as >loose B.D.)

The following further observations about bulk density provide further enlightenment about this important property:

1. A "light" or low bulk density powder builds up less compacting strength in a hopper or bin, but it has a greater inertia at rest.

2. Although a heavy powder can build up a strong compacting strength, it has a lower inertia at rest. The greater the density (or bulk density) of a mass of powder, the greater its compaction (or tensile) strength can be. Cohesion is the other possible factor contributing to the strength.

3. Fluid powders, on the average, are of medium bulk density (from 20 to 50 lb/cu ft). However, because of their fluid-like nature, there is less compacting strength build-up possible than is indicated by their bulk density.

4. Excess moisture causes a lower overall bulk density for a granular material such as sand:

Bulk density lb/cu ft

% moisture	Aerated	Packed	Working	Compaction	Flow
Sand 0.5	94	108	96	14	Free flow
Sand 6.0	62	98	75	37	Nonfree flow
Sand 8.5	63	99	75	36	Nonfree flow

The film of water causes greater cohesion and bulk and thus the bulk density is lower.

5. For a given material, the powdered form will have a lower bulk density than its granular form, because the powdered form is

bulkier and its mass takes up more volume and thus weighs less per unit volume.

To conclude the discussion on bulk density, Table 10 lists two groups of materials, one of very light and the other of very heavy bulk denisities.

TABLE 10

List of Very Light and Very Heavy Bulk Density Materials

Very light bulk material density	Density, lb/cu ft[a]	Very heavy bulk material density	Density, lb/cu ft
Carbon, activated[a]	8-20	Chromite	110-155
Carbon black	2-15	Cu and ores	110-240
Diatomaceous earth	8-18	Fe and ores	110-250
Glass fibers	3-15	Fluorspars	90-130
Magnesium carbonate, light	4-10	Iron oxides	80-180
Magnesium oxide, light	5-19	Lead	400-480
Perlite	4-10	MnO and ores	60-160
Rockwool	3- 8	Pb oxides	130-290
Silicas[b]	1-16	Pyrites	90-190
		Tin	200-300
		Tungsten carbides	530-650
		Zn and ores	130-250
Silicates[c]	2-16		
Sisal	3-15		
Stearates[d]	6-16		

[a]Ranges for different grades of the same material would also have varying bulk densities.

[b]Such as Aerosil, Cab-O-Sil, Hi-Sil, and Santo Cel.

[c]Such as Micro-Cel and Silene.

[d]Such as zinc and calcium

B. COMPRESSIBILITY

The apparent compressibility of a bulk or mass of particles is computed from the aerated and packed bulk densities of a substance. The percent compressibility of a powdered and/or granular material is:

$$\frac{\text{packed B.D. - aerated B.D.}}{\text{packed B.D. x 100}} = \% \text{ compressibility}$$

The percent compressibility will vary according to a material's (1) bulk density, (2) uniformity in size and shape (of particles), (3) hardness, (4) surface area and size of particles, (5) moisture content, (6) cohesiveness, and (7) deformability.

In actuality, compressibility is an indirect measure of the foregoing properties of a material. A very compressible nonflow material will tend to have: (1) lower bulk density, (2) nonuniform-sized particles, (3) nonuniform-shaped particles, (4) softer particles, (5) greater surface area, (6) smaller-sized particle, (7) higher moisture content, (8) greater cohesiveness, and (9) greater deforma-bility, and a substance with a lower compressibility would have the opposite tendencies.

A highly compressible material can build up a greater strength in its "bridge" or "arch." It would not be advisable, say, to vibrate a powder that has a high compressibility factor. Such powders also will tend to be very nonflowable.

The following materials have grades that are very compressible: barium carbonate, calcium carbonate, carbon black, cocoa, iron oxide pigments, kaolin clay, pigments in general, titanium dioxide, and zinc oxide. Table 11 shows the points of change in the compressibility of powdered and granular masses where the material would cause more of a problem.

TABLE 11

Points of Change in Compressibility of Powdered and Granular Masses

Flow	Problem with flow	% compressibility	Flood
Excellent	No	5-12	No
Good-fair	Minimum	13-22	No
Passable-poor	Fair	23-35	Yes
Poor-very poor	Big	36-45	No
Very poor-very, very poor	Very big	46 plus	No

C. ANGLES OF FLOW [14,20,22-28]

These include the following flow angles: repose, fall, difference, slide, spatula, internal friction, and rupture (see Fig. 1). Each angle of flow measurement has a particular unit operation wherein it is of value as a flow indicator. It would be "flow" with the unit operations of conveying, fluidization, feeding, and storage. With the other unit operations these properties, or angles of flow (except for the angle of difference) could be indicators of the ease or difficulty of mixing, blending, agglomeration, and size reduction. The lower this angle of flow is for a material, the more optimum its flow can be. With the angle of difference, it is just the opposite. Good flow is important with conveying, feeding, and storage. The agglomeration, mixing, blending, or fluidization, of a material is most readily accomplished when it has an angle of flow that is neither too high nor too low. The angles of flow are defined and discussed below.

1. Angle of Repose (ϕ')

This is the most familiar angle of flow (see a in Fig. 1). It is the angle formed by pouring a free-flowing material, or screening a nonfree-flowing powder, from a predetermined height to form a cone. The angle to the horizontal of this cone is continually measured until

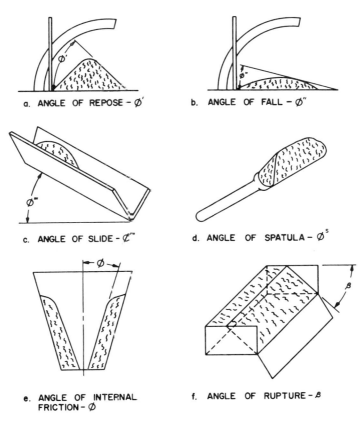

a. ANGLE OF REPOSE - ϕ'

b. ANGLE OF FALL - ϕ''

c. ANGLE OF SLIDE - ϕ'''

d. ANGLE OF SPATULA - ϕ^s

e. ANGLE OF INTERNAL FRICTION - ϕ

f. ANGLE OF RUPTURE - β

Fig. 1. Angles of flow.

it reaches a constant value. This is a readily performed test and it is indicative of the flow and other properties of a material. One limitation to the values measured is that the span of degrees between a free-flowing material and a very cohesive non-free-flowing material is not too large. This span of degrees for over 90% of materials is from 35° for free-flowing materials to 55° for a non-free-flowing substance. Some substances have ϕ' of over 70°.

Table 12 gives the characteristics of a material measured by its angle of repose (measured as directed to a constant angle).

TABLE 12

Characteristics of a Material Measured by Its Angle of Repose

Angle of repose (ϕ')	Characteristic of material
25-35	Very free-flowing; noncohesive; for granular material
25-35	Very floodable to floodable; for fluid powders
35-45	Free-flowing; some cohesiveness; for granular material
35-45	Fluidizable powders; some cohesiveness
45-55	Non-free-flowing; cohesive
55-65	Very nonfree-flowing; very cohesive
65-75	Very, very, nonfree-flowing; very, very cohesive; bulky

A high angle of repose would indicate opposite tendencies. The angle of repose has been a confusing property, for different fields of engineering have given other angle properties this name. For example, the angle of slide has been termed the angle of repose [24]. There are three other angles of flow that are similar to the angle of repose.

Angle of internal friction ϕ (see (e), Fig. 1)

Angle of rupture β (see (f), Fig. 1)

Angle of spatula ϕ^S (see (d), Fig. 1)

The angle of internal friction is the dynamic equilibrium reached between the moving particles of a material within the bulk solid as in a hopper. Generally it is a higher figure for a given material than that material's angle of repose. A distinction between these two angles of flow is: ϕ' is a measure of static equilibrium reached between the particles in a cone or heap which is uncontained. ϕ is the picture of the dynamic equilibrium reached between the particles of a bulk powder moving over the powder's stationary massed portion in containment.

The main difference in the measurements is the absence or presence
of containment. Both angles are a measure of the cohesiveness of
particles when powdered materials are measured. For granular materials,
the angles are a measure of the uniformity of size and shape of the
granules. Free-flowing materials with a low ϕ' will also have an ϕ
of similar value in many cases.

Jenike in his work mentions four angles of friction [28]:
(1) Kinematic angle of friction (ϕ'); the angle between the solid
and a wall material. (2) Static angle of internal friction (ϕ_t);
angle between the solid at an exposed surface of a pipe (flow). (3)
Kinematic angle of internal friction (ϕ); the angle between the yield
locus and axis v (x axis). (4) Effective angle of friction (δ);
the angle between the effective yield locus and the δ axis (x axis).
These angles can be determined by direct shear tests on a consolidated
material.

The angle of internal friction that is measured as a result of a
shear test is, in reality, an indirect measurement. Thus one can see
that there are many interpretations of an angle of internal friction.

The effective angle of friction, for example, has low values for
fine and dry solids, while for coarse and wet solids it has high values.
For dry sand or other free-flowing solids $\phi = \delta$. The angle of internal
friction, as measured by the author, is more akin to Jenike's static
angle of internal friction or the effective angle of internal friction.
ϕ as measured by the author is generally higher than the ϕ' of a
material. There are several ways to measure the ϕ more directly.
Figure 1e shows one simple way to measure ϕ. There is not close
agreement in ϕ measurements using different methods. Therefore, it
is important to relate the values obtained to the method used and to
the type of test being conducted.

In a fluidization test ϕ can be related to the L/D (length/diameter)
ratio corresponding to the onset of slugging. With the use of ϕ, a
prediction can be made concerning bed heights that can cause slugging
[25-26].

The angle of rupture is an angle of flow similar to the angle of
internal friction. It is the angle or interface formed when a bulk

of massed particles (powder) moves under the influence of gravity
over its stationary bulk phase. The values of ϕ and β can be similar
for a given material, or, they can be quite different. In general,
for a given material the angle of internal friction is the highest
value, the angle of rupture is next, and the angle of repose is the
lowest.

A simple angle of flow measurement is the angle of spatula (ϕ^S).
It is a quickly measured, relative angle of internal friction of a
material. A spatula (5 x 7/8 in.) is inserted into the dry material,
and then brought up and out. A free-flowing material forms an angle
of repose, while a non-free-flowing material forms several irregular
angles of flow on the blade. In either case, these angles of repose
are measured and an average taken. Then the spatula is gently tapped
and new angles form. All measurements are averaged to obtain ϕ^S.
Except for very free-flowing materials, ϕ^S is always a higher value
than ϕ'. For a material to be considered free-flowing, its ϕ^S must
be less than 40°. If ϕ^S exceeds 40°, the material is not considered
free-flowing.

The ϕ, β, and ϕ^S, like the ϕ', will characterize a material
similarly as their values go lower or higher.

All the foregoing angles of flow are an indirect measure of cohesion, surface area, size, shape, uniformity, fluidity, bulk density,
porosity, and deformability. They are a direct indication of the
flowability of a substance "en masse."

2. Angle of Fall (ϕ'')

The angle of fall is the new angle of repose assumed by a material
after its original ϕ' is "collapsed" by dropping a steel bushing, from
a specified height, 5 times on the ϕ' indicator plate (see Fig. 1b).
The ϕ'' is important in indicating how floodable or how fluidizable a
material is [20]. The lower ϕ'', the more fluidizable a mass of
powdered particles. For granular materials, the lower ϕ'', the more
free-flowing the material. The ϕ'' of a substance is an indirect

indication, of its fluidity, shape, size, uniformity, entrapped air,
and cohesiveness. It is a direct measure of the relative flow and
stability of a dry material.

3. Angle of Difference (ϕ^D)

This is the difference between the angle of repose and the angle
of fall for a given material. It also indicates the potential floodability or fluidity of a substance. The angle of difference gives a
different picture or measure of a material's fluid-like nature. A
material could have a high ϕ'' which indicates a less of a fluid-like
nature, but also have a high ϕ^D which indicates more of a fluid-like
nature. Also, for granular substances a high ϕ^D indicates more of a
free-flowing nature. The greater ϕ^D, the more flowable or the more
floodable the material.

4. Angle of Slide (ϕ''')

This angle of flow has been characterized as the angle of repose
in some publications [24] (see Fig. 1c). It is a measure of the
adhesiveness of a mass of particles to a dissimilar surface. It is
measured by observing the angle at which the bulk material will slide
down a prime-coated piece of steel. The value varies according to
films such as dust, moisture, temperature, surface finish, and oxide.
The angle of slide is an important property to consider in the design
of hoppers and conveyors. The more free-flowing the material, the
lower its angle of slide.

D. DISPERSIBILITY

The floodability (or fluidity), dustiness, and dispersibility of
a material are interrelated. The more dispersible the powder, the
more dusty and floodable (or fluid-like) is its mass of particles.

Dispersibility is measured by dropping a specified weight, e.g.,
10 g, of a powder through a 4 in. plastic cylinder onto a 4 in.

watchglass [20]. The powder left on the watchglass is weighed. In
tests performed on over 3,520 dry materials, the five most floodable
materials left only 50% of their original 10 g on the watchglass.
This 50% was taken as the standard for a very floodable or fluid-like
material. This test is a practical demonstration of the ability of
a mass of powdered particles to flood or to be fluidized. One variable
is the equilibrium moisture content of the material. Very light
materials lose most of their mass through dustiness and not because
of an unstable fluid-like nature. Dispersibility is a measure of the
cohesiveness, particle size and surface area, density, bulkiness, and
moisture content of the substance.

E. COHESION, APPARENT SURFACE OR APPARENT SURFACE COHESION [30]

The term cohesion has meaning in more than one area of science.
It is called by physicists a force holding together the atoms in a
molecule. Here, however, the colloidal- to granule-size level is of
concern, rather than the atomic level. Here it is the apparent cohesive forces that are effective on the surfaces of these larger-than-
molecule-sized particles that need to be measured. Dry, granule-size
(and larger) particles do not have enough effective cohesive forces.
Granules are large, dense units with many particles whose forces are
mostly satisfied. The forces that are not balanced out are too few
to be of any consequence because the surface area is overbalanced by
the larger particle size. If the granule is broken up into many
powdered units there will be a relatively greater surface area. The
ratio of size to surface area has reversed. More moisture molecules
and impurities (as atoms) can be adsorbed. Broken bonds will be
exposed. The particle will be more deformable. Van der Waals and
London forces will also be exposed in greater number. All these factors
cause a greater attraction and bonding between particles, which is
termed apparent cohesion. Because of their greater deformability,
softer particles have more effective cohesive forces.

Cohesion is a fundamental flow property. Its evaluation is an
important factor in predicting flow. It is a direct measure of the
amount of strength a mass of particles can build up in a hopper or

bin. Many of the other property measurements are indirectly a measure of this surface cohesive force. A high cohesion will make fluidization, feeding, drying, storage, and classification difficult.

F. SIZE DISTRIBUTION, MESH SIZE, UNIFORMITY COEFFICIENT

In 1968 Cheng reported that particle-size distribution is a major factor that determines the tensile strength of a powder [29]. Other factors are density and cohesion. He states that the higher the density of a powder compact relative to particle density, the greater its tensile strength. We have already stated that the higher the bulk density, the greater the strength possible. A nonuniform distribution of particle size would lead to a greater "cementing" as the smaller units cling to the larger particles and so cause a continuous cohesive force to be transmitted throughout the mass.

The particle-size distribution is determined by making a screen or sieve analysis of the material. The screen analysis results point out:

1. The powder and granule fractions of a material or the ratio of floodable and/or archable powder to the free-flowing granule units (given roughly in Table 13).

TABLE 13

Approximate Powder Fraction and Granule Fraction
Resulting from a Screen Analysis

Bulk density (lb/cu ft)	Powder Fraction (mesh size)		Granule fraction (mesh size)	
10-20	-100 and larger		+100 and larger	
25-55	-200	" "	+200	" "
(average material)				
60-90	-325	" "	+325	" "
125-200	-400	" "	+400	" "
Polymers	-325	" "	+325	" "
(average wt.)				

2. The uniformity in size distribution. The more uniform, the
more flowable and vice versa. As already stated, a nonuniform size
distribution leads to a higher tensile strength in a powder or powder-
granule mass.

With granular materials there is no effective cohesive force that
can be measured, thus instead of a cohesion test, a uniformity coefficient is calculated, based on the screen analysis of the material.
In other words, the uniformity coefficient is an alternate criterion
of flow. The uniformity coefficient is the numerical value arrived
at by dividing the sieve opening that will just pass 60% of the
material by the sieve opening which will just pass 10% of the sample.
As already stated, the higher this coefficient, the more nonflowable
and vice versa.

Table 3 lists some typical sizes of some typical industrial materials. Notice the relationship of size to surface area.

Table 14 indicates the relationship of mesh size to particle size,
given both in micrometers and in the more familiar inches.

The mesh analysis is an important test to make prior to any of
the unit operations. As mentioned early in this chapter, surface
area is the most important property of a material and size is directly
related to surface area. With a mesh analysis study, one can determine
the possible difficulties to be encountered in handling or processing
materials in any unit operation. The mesh analysis can pinpoint possible trouble and it also gives the uniformity coefficient of the material.

G. SUMMATION

The most important characteristics and properties of particles
en masse have been enumerated. A combination of these properties can
be used to evaluate the flow of materials en masse [20].

IV. BULK-SOLID MECHANICS [22-25,27,28,39,44]

Many of the properties of particles that have been discussed are
mechanical properties. It follows now that the mechanics of bulk

TABLE 14

Relationship of Mesh Size to Particle Size

Sieve or Mesh No.[a]		Size of openings					
		U.S.			Tyler		
U.S.	Tyler	Micrometers	Millimeters	Inches (approx.)	Micrometers	Millimeters	Inches (approx.)
3	3	6,350	6.35	0.250 (1/4)	6,680	6.68	0.263
4	4	4,760	4.76	0.187 (3/16)	4,699	4.70	0.185
6	6	3,360	3.36	0.132 (1/8+)	3,327	3.33	0.131
8	8	2,380	2.38	0.094 (3/32)	2,362	2.36	0.093
10	10	2,000	2.00	0.079 (5/64)	1,651	1.65	0.065 (1/16)
14	14	1,410	1.41	0.056	1,168	1.17	0.046
20	20	840	0.84	0.033 (1/32)	833	0.833	0.033 (1/32)
-	28	-	-	-	589	0.589	0.023
30	-	590	0.59	0.023	-	-	-
35	35	500	0.50	0.020	417	0.417	0.016 (1/64)
40	-	420	0.42	0.016 (1/64)	-	-	-
-	48	-	-	-	295	0.295	0.012
60	-	250	0.25	0.010	-	-	-
-	65	-	-	-	208	0.208	0.008
100	100	149	0.149	0.006	147	0.147	0.006
140	-	105	0.105	0.004	-	-	-
-	150	-	-	-	104	0.104	0.004
200	200	74	0.074	0.003	74	0.074	0.003
270	270	53	0.053	0.002	53	0.053	0.002
325	325	44	0.044	0.0017	43	0.043	0.0017
400	-	37	0.037	0.0015	-	-	-
625	-	20	0.020	0.0008	-	-	-
1,250	-	10	0.010	0.0004	-	-	-
2,500	-	5	0.005	0.0002	-	-	-
5,000	-	2.5	0.0025	-	-	-	-
12,500	-	1.	0.001	-	-	-	-

[a] Sieve or mesh numbers of 100 indicate that there are 100 meshes per linear in. or 10,000 per sq. in., or 6 mesh indicates 36 meshes per sq. in. [b] 1 micrometer = 0.00004 in.

solids be probed. The discussion will include shear tests, which are
much used in soil science.

The mechanical properties of bulk solids are in reality a direct
result of the merging properties of the individual particles within
the bulk solid. Again, the most important merging factor is the
surface area of the individual particles which, when multiplied by
the many particles in a massed solid will, without doubt, contribute
to the basic bulk-solid properties and its mechanics of flow. The
magnitude and the effectiveness of the surface forces which include
van der Waals-London forces, surface energy, hygroscopicity, electrical
force, and the adsorbed impurities, depend entirely on the relative
magnitude of the surface area. A large surface area yields high angles
of internal friction, high shear forces, and a high cohesion; while a
small surface area yields small angles of friction, shear forces, and
cohesion.

A mass of fine particles can be thought of as a continuous mass
in which normal compressive and tangential stresses due to shear can
be maintained. Properties of the massed particles are the same in all
directions.

The measurement of the shear forces and the angle of internal
friction is directly influenced by the bulk properties of bulk density
and compressibility, cohesion, and mesh-size distribution. A high
bulk density, a wide compressibility, a high cohesion, and a wide
mesh-size distribution lead to a high shear force and angle of internal
friction. A mass of large granule-size particles actually can transmit
its force, or stress against hopper sides and opening because of the
particles free flow. But a mass of very fine, cohesive, powder particles, because of its nonflow, does not have such a continuous thrust
or force (or stress) against the hopper sides or opening. Its stresses
are more within its bulk mass. Only when this inertia is overcome is
the force of the material felt against the hopper walls and opening.

In addition to its importance for determining flow properties,
bulk-solid mechanics are also important in other unit operations. A
powder material with a high shear result will show nonflow qualities.
This same quality will also be critical for conveying, fluidization,

sintering, and storage; and noncritical for agglomeration, size reduction (in some cases), mixing, and drying.

One method of determining stresses in bulk solids is by using the
Mohr circle of stresses, a graphical procedure for evaluating the
equilibrium of stresses within an "arch" or "bridge" of the massed
bulk solids in a hopper, under no-flow situations. Generally a Mohr
circle is drawn tangential to a curve plotted from several shear-
normal stress determinations. This semicircle intersects the normal
stress axis at the two principal normal stresses, major and minor.
If a Mohr circle is drawn tangential to each of several shear-stress
points, then a curve is drawn tangent to these several circles to the
axis (y). This "envelope" along with the Mohr semicircles graphically
portrays the stresses and nonflow characteristics of a material. Major
and minor stresses are pinpointed, a cohesion factor is arrived at,
and a minimal hopper opening can be calculated to prevent arching.
Figure 2 shows graphically a Mohr circle shear-stress determination.
Circle B goes through the origin at zero, which here is the minor
normal pressure. This is the stress equilibrium in the arch or bridge
of material at nonflow conditions. δ_2 of Circle B can be used to
determine a theoretical minimum hopper opening [28,39].

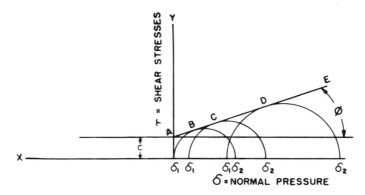

Fig. 2. Normal pressure versus shear stresses. δ_1 = minor
normal pressure, δ_2 = major normal pressure, ϕ = angle of internal
friction, c = cohesion, ABCDE = shear-stress curve.

There are several methods in use to determine shear stresses of consolidated materials. Each procedure can be consulted and then a method can be devised which best relates a given shear stress to the problem at hand. The following two shear tests are in use in soil science studies:

1. Triaxial shear. A lateral pressure held constant while a compressive force loads the massed bulk solids to failure. A curve is developed.

2. Direct shear. A horizontal shear load is applied while a vertical load holds a constant pressure. Again a curve is developed. Jenike developed his own direct-shear test method [28] because he found that the direct-shear unit used in soil science gave cohesion values outside the useful range needed for bulk solids. He further states that soils under their usual degree of consolidation do not flow by gravity in any storage system.

To summarize the following properties are of great importance in the different test procedures and also in bulk-solid characterization:

1. Shear stress. A measure of the friction between particles or the cohesiveness of a bulk material. The important variable factor here is the amount of normal pressure or consolidating pressure used.

2. Angles of internal friction. Theoretical angle formed between a shear stress-normal pressure curve and a normal pressure abscissa. Jenike gives four angles of internal friction [28]. Theoretical angles of internal friction seem to be of a lower order of magnitude than a directly measured angle.

3. Cohesion. A theoretical value measured on the y axis (shear stress axis), between a shear-normal pressure curve and the x axis (normal pressure axis). Actual cohesiveness of a bulk material seems to be of a higher order of magnitude than this measured theoretical value.

4. Bulk density, compressibility (consolidation). An important measure of strength that can be built up in a bulk solid. Bulk density

is used in shear-stress equations. The important factor here is what bulk density figure should be used in order to give a true picture of the strength built up in the solid mass of particles.

In performing shear tests, the engineer should use consolidating pressures and bulk density that relate directly to the problem at hand. The use of shear test results along with correctly designed-for-flow hoppers (or bins) which have adequate-sized openings assures the continual flow of archable materials that have some free-flow characteristics. Materials that have no free-flow characteristics or are nonflow masses need, in addition, some form of flow inducement or aid.

V. CLASSIFYING BULK SOLIDS OR MASSES OF PARTICLES [31,42]

After testing and evaluating the flow of over 3,000 materials, a study was made of their different properties. From this study, a classification system for masses of particles was derived [31]. Substances are classified into three main groups: Corpuscular, laminar, and fibrillar. These three groups are further broken down into subgroups according to particle size, and whether granular or powdered. This classification system is based on the physical properties of materials. A system based in part on the chemical properties of materials is in preparation, however, the physical classification system can be very useful for a unit-operation engineer in classifying material. The procedure would consist of: (1) A mesh analysis-uniformity coefficient, (2) microscopic or visual examination for coarse materials, and (3) bulk density determination to establish granule-powder borderline. From the data obtained, the engineer can place his material in the system and then read the average properties of flow. The physical classification system is given in Table 15 [31]. Depending on what unit operation is involved, the engineer can relate the properties of his material to the optimum particle properties required. A detailed interpretation of Table 15 is given in reference 31.

TABLE 15

Classification — Average, Relative Properties of Flow of Masses of Particles[a]

Class	Uniformity coefficient[c]	Powdered fraction (%)[c]	Flow (%)	Flood (%)	Arch-ability[d]
I. Corpuscular					
A. Granules					
1. Uniform	~1-10	~5	70-100	0	No
2. Nonuniform	15-30	"	60- 75	0	MHUIP
3. Very nonuniform	30+	"	50- 70	0	HUIP
4. Damp	30+	"	30- 65	0	HU
5. Soft (or sticky)	-	"	40- 75	0	MHUIP
B. Powdered granules					
1. Uniform	1-12	<30	70- 80	0	No
2. Less uniform	8-18	"	55- 70	0	MIP
3. Nonuniform	15-30	"	50- 65	0	Y or HU
4. Very nonuniform	30+	"	30- 60	0	HU
5. Damp	30+	"	40- 65	0	Y
6. Soft (or sticky)	-	"	50- 70	0	MIP
C. Fluid powders					
1. Very fluid a. Granular	-	60- 90	50- 70	70-90	IVEP
b. Granular and powdered					
c. Powder		30- 60	50- 65	30-70	IEP
		95-100	45- 65	80-95	IP
2. Fluid a. Granular	-	60- 90	45- 55	50-75	Y or IP
b. Granular and powdered					
c. Powder		30- 60	40- 60	20-60	Y or IP
		95-100	35- 50	45-75	Y or IP
3. Fluid cohesive a. Granular	-	60- 90	20- 45	25-60	Y
b. Granular and powdered					
c. Powder		30- 60	20- 40	10-35	Y
		95-100	10- 40	15-45	Y or VG
D. Cohesive powders					
1. Powder	-	95-100	5- 25	0-20	VG(P)
2. Granular and powdered	-	30- 60	5- 40	0-20	Y or VG

			Size[b]	[c]	[d]	[d]
II. Laminar						
A. Micaceous	1.	Thin	1/2 to 3 in.	1- 15	0	HUVG
	2.	Thick	1/4 to 1 in.	10- 20	0	HU
	3.	Powdered	-200	30- 50	35-60	IP
B. Film	1.	Very thin	1/16 to 1/2 in.	1- 10	0	HUVG
C. Chips	1.	Fine, uniform	+100 to +10	30- 80	0	No
	2.	Fine, nonuniform	- 60 to +10	60- 70	0	MHUIP
	3.	Large, nonuniform	- 10 to 3 in.	30- 40	0	HU
D. Flakes	1.	Thin	-100 to 1 in.	1- 20	0	HUVG
	2.	Fine, uniform	- 20 to 3/8 in.	30- 80	0	No
	3.	Fine, nonuniform	- 40 to 3/4 in.	45- 50	0	HUIP
	4.	Powdered	-200 to 1/4 in.	50- 70	0	HHIP
III. Fibrillar						
A. Stume	1.	Very short	1/4 to 3/8 in.	50- 60	0	HUIP
	2.	Short	1/4 to 1-in.	10- 20	0	VG
	3.	Long	2 to 3 in.	1- 10	0	HUVG
B. Fibrous	1.	Bunches	1/2 to 2 in.	5- 20	0	HUVG
	2.	Fine bunches	-100	20- 30	0	HU
	3.	Powder, coarse	-200 to +10	45- 55	0	IP
C. Acicular	1.	Fine	+100 to +40	65- 75	0	MHU
	2.	Fine, nonuniform	-100 to +10	55- 70	0	HUIP
	3.	Medium	+ 20 to 7/8 in.	30- 40	0	HU
	4.	Powder fluid	-200	35- 50	40-60	IP

[a] Excerpted by special permission from Chem. Eng., Feb. 1, 1965, p. 70.

[b] For laminar and fibrillar materials the average size is given in inches and/or mesh size.

[c] For the average weight material powdered fraction is the -200-mesh fraction; for heavier materials it is -325 or finer; for lighter materials it is -100 or coarser.

[d] M = may, U = up, P = packed, E = excess, G = greatly, H = hang, I = if, Y = yes, V = very.

57

A. FLOWABILITY, STABLE FLOW

A free-flowing material tends to flow steadily and consistently, as individual units, even through a fine orifice (i.e., large enough for the size granules). It is not subject to arching in a hopper or flooding from or through a hopper. The flow of material is stable, under control, and constant at all times with no aid.

A non-free-flowing material does not flow consistently and steadily, and is subject to arching or doming in a hopper or bin and will need aid and/or effectual hopper design to keep it flowing. The particles here will tend to flow "en masse." Powders and powdered granular materials are examples of non-free-flowing materials. Materials consisting of long fibers are also of a very nonflow nature. Examples of non-free-flow materials are pigments, diatomaceous earths, asbestos, cellulose, fiberglass, barium carbonate, mica (thin), and sisal.

The flowability of materials is important in the unit operations of conveying, feeders, agglomeration, granulation, additives, and storage.

B. FLOODABILITY, FLUIDIZABILITY, UNSTABLE FLOWABLE POWDERS

Floodable flow is unstable, liquid-like flow. It is a gushing, uncontrollable, spattering flow. A floodable material has the ability to be excessively aerated under either of two conditions:

(1). The sudden breaking of an arch or bridge of material in a hopper. Floodable or fluid-like materials, by their very less-cohesive nature, build up less strength in their "bridge" and thus can readily fall through the hopper from this weak arch, or "hang up."

(2). The flowing into and through a hopper which had been partially or entirely empty. Under these conditions, the floodable material will entrain air as it falls, assume a lower bulk density, and take up a greater volume; thus it will flow like a liquid — spattering and gushing.

Very floodable and thus readily fluidizable materials include
activated carbon, acrawax, alumina powder, aluminum powder, clays
(some more than others), coal powder, fly ash, lime (hydrated, not
air slaked), phosphate rock powder, phenolic resin, starch (tapioca),
and talc.

As mentioned previously, bulk properties of a material can be
used to evaluate its flow. Flowability can be determined or evaluated
by the use of the following bulk properties [20]: angle of repose
(25 points), compressibility (25 points), angle of spatula (25 points),
and cohesion (powders) and uniformity coefficient (granular) (25-point
total for both).

Points are assigned to a material according to its flow properties.
After testing over 3,000 materials, the points of change in values
for each flow property in going from free-flow to non-free-flow have
been established, as related in Table 16.

A flowability below 20% is a danger signal, requiring special
attention or special engineering for a feed system. Reference 20
has more details.

TABLE 16
Relation Between Type of Flow and Flowability

Type flow	% flowability	Hopper agitation
Excellent	91-100	None
Good	81- 90	None
Fair	71- 80	None
Passable	61- 70	Possibly, could vibrate
Poor	41- 60	Yes
Very poor	21- 40	Extra or special agitation
Very, very poor	0- 20	Special agitation or special engineering

The following factors are used in arriving at a floodability percent: flowability (25 points), angle of fall (25 points), angle of difference (25 points), dispersibility (25 points).

Floodability can be evaluated the same as flowability. Table 17 translates points to type flooding, etc. a floodability percent of over 80 indicates that special attention be given to a very fluid-like material. Again, see reference 20 for more detail.

C. ARCHABILITY [27-44]

The arching or bridging of a nonflow material in a hopper is all too common an occurrence in feeder systems. Free-flowing particles tend to obey the pull of inertia and flow downward more consistently to the hopper exit. Nonflow materials drop or collapse "en masse" as agglomerates toward the hopper exit. There are forces of compression due to the sides of the hopper and the weight of the material. If a greater strength develops in the arch (greater than stresses at hopper opening), then doming (or arching) occurs over the arching. In other words, the granular powder or powder material has enough strength in its consolidated mass to support its own weight. "Hang

TABLE 17

Relation Between Type of Flooding and Floodability

Type of flooding	% floodability	Rotor or a flood control
Very floodable	80-100	Necessary, positive seal
Floodable	60- 79	Necessary
Inclined to become aerated or fluidized	40- 59	Desirable
Could become aerated or fluidized	25- 40	Depends on velocity of fall or rate of feed
Will not flood or become fluidized	0- 24	None

up" is a term used to describe what could be called a weak arch or a
momentary arch within a confining area. It would take only a small
amount of energy to cause this "hang up" to fall from the hopper.
Fluid-like powders, granular powders, powdered granules, some mica-
ceous particles, and acicular crystals (not too long) are among
materials that form a weaker arch, i.e., are classifiable as "hang
up" materials. In general, these "hang up" materials have relatively
lower compressibilities, between 20 to 30%. Fluid materials can
build up a weak arch or 'hang up" within a hopper, and when this weak
arch breaks, the material floods through the feeder. Fluid-like
particles are less cohesive because of a film of air around each
particle and they cannot be extensively compressed.

Long-fibered materials or materials with long acicular units will
tend to "hang up" to a large extent, however, and special engineering
is needed to move them.

The critical factors in hopper design are the slope of the sides
and the hopper gate or opening at its base. Following are three
equations which can be used by the design engineer as a guide for the
size hopper opening needed. If the opening is too small, nonflow
conditions for a nonflow material are present, with a resultant arch
or bridge or "rat-hole."

Smith [36] gives an equation stressing static base pressure at
hopper opening (hopper's side inclination $>5°$):

$$P_b = \left[\frac{M}{3-B}\right]\left[\frac{(R_b + Z \tan \psi)^{(3-B)} - R_b^{(3-B)}}{R_b^{(2-B)}}\right], \qquad (3)$$

where

$$B = [2/\tan \psi][(1 - K') \sin^2 \phi + K'][\mu' + \tan \phi]$$

and $M = \rho/\tan \psi$, P_b = base pressure (lb/sq ft), R_b = radius of base
of hopper opening (ft), ψ = angle of cone of supported solids with
the vertical, Z = vertical distance from top of solid mass (ft), and
ρ = bulk density of material (lb/cu ft). M and B are constants as

defined above and ϕ = angle of hopper wall with vertical, K' = ratio
of normal to applied pressure in solid mass, and μ' = coefficient of
friction, solids to bin wall. Equation (3) is theoretical, and experimental data are needed to substantiate the mathematics. Smith
states:

(1) K' should be small, say, 0.1 or 0.2; and (2) discharge opening
should be large enough to give a considerably greater pressure. A
minimum pressure value of 50 lb/sq ft is probably correct in most
cases.

Stepanoff gives the following formula for the width, a, of a
rectangular slot hopper opening [27]:

$$a = \frac{2\tau_0(1 + \sin\phi)}{\gamma} , \qquad\qquad (4)$$

where a = width of slot, τ_0 = shear stress (initial), ϕ = angle of
friction between yield locus and abscissa, and γ = bulk density.

For a round opening,

$$d = \frac{4\tau_0(1 + \sin\phi)}{\gamma} \qquad\qquad (5)$$

where d = diameter of opening. a and d are the maximum dimensions
of a hopper orifice that still permit formation of an arch.

Jenike gives Eq. (6) for a hopper outlet dimension B for either
width of a rectangular outlet, or side of a square outlet, or diameter
of a circular outlet. This outlet B will be sufficiently large to
have, for doming [28],

$$B = H(\theta')V_1A\gamma \qquad\qquad (6)$$

where B = dimension for hopper outlet (ft), H = moisture content (%),
θ' = hopper shape measured from the vertical, V = normal force applied
to a shear cell during shear (lb), A = area of a shear cell (1/13 sq
ft), and γ = bulk density of solid (lb/cu ft). Or the critical diam-

eter B for no-doming over a circular outlet is given approximately
by

$$B = 2.2F/\gamma \tag{7}$$

where B = diameter of hopper outlet, F = strength of solid, and γ
= bulk density (lb/cu ft).
Also,

$$B = \frac{F_c(1 + m)}{\gamma} \tag{8}$$

where F_c = 1 for strongest possible arch, m = 0 when opening is a
slot of width B, or m = 1 when opening is circular with diameter B,
F_c = unconfined yield strength, and γ = bulk density.

These theoretical formulas can be of use in determining hopper
openings for materials that "hang up" or that do have some free-flowing
characteristics. Such outlets can, in these cases, insure continuous
flow. For highly compressive, cohesive no-flow materials, these
theoretically computed hopper openings help to minimize doming or
arching over the hopper outlet, and with some positive aids, such as
agitators, vibrators, or baffles, continuous flow may be assured.

VI. MASS OF PARTICLES, CHEMICAL ASPECTS [14,21,31,45-50]

The subject of arching concerns the mechanical or physical aspects
of materials. The subject massing of particles involves their chemistry.
Many softer, more deformable and hygroscopic crystals have a tendency
to mass and/or lump or "cake" in a bin or hopper under a high pressure
and exposed to a changing temperature and relative humidity. There
is a natural tendency for crystals or particles to decrease their
overall surface area, or, in other words, their free surface energy.
This natural sintering goes on until the particles reach a size where

the free surface energy is neutralized out, which is critical with
more deformable particles.

Among materials used as flow-conditioning, antimassing, or anticaking agents are:

Acid magenta (dye)	Kaolin clay
Aerosil (silica)	Magnesium carbonate
Armeen T (cationic amine)	Microcel (calcium silicate)
Armact (cationic amine)	Phosphate rock
Cab-O-Sil (silica)	Hi-Sil 233 (hydrated silica)
Calcium silicate	Quso M51 (hydrophilic silica)
Calcium sodium silicoaluminate	Santocel (silica)
Cornstarch	Silene (calcium silicate)
Diatomaceous earth (silica)	Tricalcium phosphate
Fullers earth (clay)	Vermiculate
Guar gums	YBS (yellow prussiate soda)
Cellulose gums	Zeolex (sodium silicoaluminate)

The use of the foregoing conditioning agents does not always
produce 100 percent results, but they are certainly of great use in
many applications where the massing of particles is a very critical
impasse to flow. There is an optimum additive and an optimum amount
of conditioner for each application. Amounts added will range from
1/2 to 10%. In most cases from 1/2 to 1% suffices. Too much of some
additives could cause a flooding problem. In some cases, while flow
may not be greatly improved, the tendency to mass is greatly lessened.
With very fine cohesive powders like barium carbonate, flow is not
improved very much, but enough so that a standard dry feeder with
agitators can work more effectively.

Table 18 gives in compact form the results of using flow conditioning agents on five different nonflow materials. A study of their
flow before and after addition of the agent shows this. The barium
carbonate particles took the highest amount of additive because of
their fineness. In summary, to deal with arching and massing the
engineer must: (1) know and understand the properties of the material

TABLE 18

The Use of Flow Conditioning and Antimassing Additives

Material	Cakes or lumps	Angle of repose	Flow	Floodable	Archable	Cohesion(%)
Flurometrone (herbicide)	Yes (lumps)	47	Very poor	No	Very	40
Flurometrone + 1/2% Cab-O-Sil M5[a]	No	45	Poor	Yes	If packed	10
Flurometrone[a] + 1% Cab-O-Sil M5[a]	No	40	Passable	Very	If very packed	5
Barium carbonate	-	57	Very, very poor	No	Very	85
Barium carbonate + 1% Cab-O-Sil[a]	-	57	Very poor	May	Very	65
Barium carbonate[a] + 8% Hi-Sil[a]	-	55	Very poor	Yes	Yes	50
Barium stearate[a,b]	-	50	Poor	Very	If very packed	15
Barium stearate[a,b] Silene	-	50	Poor	Very	If very packed	15
Salt	Cakes greatly	45	Very poor	May	Yes	-
Salt + Guar gum (50% = -325 mesh)[c]	Slight tendency to cake	30	Passable	Yes	If very packed	-
Salt + 1% Silene (75% = -200 mesh)[c]	Tendency to cake	43	Poor	Yes	If packed	-

[a] Additive as flow conditioner. [b] Additive as antimassing additive. [c] Additive as anticaking additive.

at hand, (2) design storage and flow system around this material, and
(3) try conditioning agents if possible or practical.

VII. APPLICATION

The relationship or application of particle properties to unit
operation has been touched on in several areas already (see Table 1).
What follows is a summary of this relationship "en masse."

A. SIZE REDUCTION [51-55,81]

The important factors or properties in crushing and grinding
particles are: abrasion, size, moisture content, flammability, explosiveness, temperature limitations, toxicity, and chemical makeup.
Abrasiveness or toughness is the most important particle property to
consider in size reduction. The knowledge of a particle's hardness
can be used in determining its grindability index. Some soft materials
can also cause difficulty. For example, talc and graphite are easy
to grind down to a point, but then clogging can be a problem. Some
low melting-point hydrated salts may give off water and also cause
some clogging. Very deliquescent salts such as $CaCl_2$ can also cause
a wet problem. Particle size is the next important property to have
knowledge of. One must know the size of the particle before and after
reduction. The particle moisture content (both free and bound), toxicity, stability, and explosiveness with air must also be taken into
account in order to have an efficient and safe operation. Kiln-fired
limes also can cause difficulties in size reduction, especially over-
burned and hard-burned limes.

B. CONVEYING [42,56]

The following properties of solids are critical to efficient
conveying:

Bulk density (capacity computations)

Lump size greater than 3/4 in. lump (do not use screw conveyor, etc.).

Angle of slide (whether material can be conveyed on belt conveyor).

Flowability (ease of conveyance).

Floodability (difficulty of conveyance).

Toxicity or explosiveness of dusts (safety of conveyance).

Shape, irregular or fibrous (difficulty of conveying).

Particle size (problem with some conveyors).

Corrosivity (materials of construction).

Degradability (critical with some conveyors).

Contaminability (not ideal to use conveyors).

With explosive dusts, conveyors must be completely covered and explosion doors provided.

C. PNEUMATIC CONVEYING [57-62]

The following properties are important to pneumatic conveying:

Packed bulk density (power and air requirements).

Aerated bulk density (feeder and hopper capacities).

Particle sizes (dust collector needs, types of seals, feeder needs, minimum conveying velocity).

Hardness (type of system, material of construction, bearing needs).

Hygroscopicity (need of flow inducers, type of feeders, air drying needs).

pH and corrosiveness (material of construction, air drying needs).

Cohesiveness (type of system, air drying needs).

Floodability (type of system, type of flow inducers in hopper, deaeration requirements).

Angle of repose (hopper design, type of flow inducers in hopper).

Toxicity (type of dust collector, venting system in hopper).

The most important property for which to design safety features
in the pneumatic conveying system is the stability of a material's
dust or its possible explosiveness with air. The following procedure
should be followed:

1. Eliminate possible ignition sites such as areas of impingement,
friction, and electrostatic discharge.

2. Check materials to be handled for their ignition sensitivity,
explosion severity index, and explosability index. The minimum and
range of dust concentration in the air is also critical. (Table 7
lists most of the explosive dusts.)

3. Eliminate ignition sites by changing metal or reducing drops
and minimize an explosive situation by changing rates and/or reducing
the dust.

D. FLUIDIZATION [25,62-65]

Pneumatic transport (used for large particles), where the entire bed is entrapped within the gas phase, has just been discussed.
We now consider a bed of particles of about 10 to 100 μm that is being supported in a fluid state by the use of a gas, such as air.
This is called dense-phase fluidization. The most important solids
characteristic here is particle size. Ten-μm-size units exhibit
cohesive flow, 60-μm-size units aggregate flow, and 100-μm-size units
slug flow. The preferred sizes for fluidization are:

μm	Volume %
0-20	5-25
20-80	30-85
>80 (but fine)	5-35

Other critical properties are:

Flowability (the more flowable a fine powder, the more fluid-like
and the more readily fluidized).

Hardness (possibility of wear in system).

Hygroscopicity (moisture works against fluidization).

Explosiveness of dusts (See Sec. C, Pneumatic Conveying, regarding dusts explosive with air).

Dispersibility (the more dispersible, the more readily fluidized).

Cohesiveness (the less the cohesion, the more readily fluidized).

Compressibility (the less the compressibility of a fine powder the more fluidizable).

E. FEEDERS [66-70]

The critical or important particle properties germane to a good feeder installation are:

Flowability (correct type feeder with accessories, if needed).

Floodability (need of rotor, seals).

Hygroscopicity (% relative humidity that is critical to efficient flow. A 40 to 50% humidity is more ideal than 60 to 90%.

Corrosivity (need of resistant materials especially if material is excessively hygroscopic or damp).

Angle of repose (to evaluate flowability).

Compressibility (to evaluate flowability).

Angle of spatula (to evaluate flowability).

Cohesion (to evaluate flowability).

Uniformity coefficient (to evaluate flowability).

Flowability (to evaluate floodability).

Angle of fall (to evaluate floodability).

Angle of difference (to evaluate floodability).

Dispersibility (to evaluate floodability).

Explosiveness of dusts (need of explosion-proof motors, etc.).

Size of particles (indicative of flowability and floodability).

Bulk density (capacity computations).

F. MIXING AND BLENDING [71,72]

Examples of solid-solid blending are:

Stucco mixes	Rubber
Refractories	Assorted nuts

Molten glass	Fertilizers
Drink mixes	Pills
Baking mixes	Dry cells

The important properties here are:

Bulk density (heavy particles seek bottom).

Particle density (heavy particles seek bottom).

Particle size (smaller particles seek bottom).

Particle shape (round or smooth particles seek bottom).

Floodability (fluid particles aerate, have cushion of air around particles).

Electrostatic forces (repulsion to blending).

Cohesion (hard to intimately blend if cohesion is high).

Flowability (powder with some flow more readily mixed than no-flow powders).

G. AGGLOMERATION [72-75,84]

Agglomeration as a general term meaning size enlargement includes tableting, briquetting, agitation, extrusion, sintering, and heat reaction as subprocesses. Other names for agglomerated units are tablets, pellets, flakes, spheroids, spaghetti-like particles (extrusion), pillows, and granules. The properties next mentioned, in all liklihood, pertain to all such agglomerated units. Among the materials agglomerated are fertilizers, ores, pharmaceuticals, food, and catalysts.

The properties of solids that are important to agglomeration are:

Flowability and cohesiveness (lubricants and binders can impart these characteristics, for compaction).

Particle size (too fine a particle means high cohesion, causing poor flow).

Flowability (for processing in pellet mill and other units).

Surface forces (important to agglomeration for strength).

Adhesiveness (agglomerate should not stick to punch faces).

Hardness (too hard a particle difficult to agglomerate).

Particle size distribution (enough fines to cement larger particles together for a stronger unit).

Lubricants, binders, and plasticizers are used to insure optimum characteristics for agglomeration.

H. SINTERING [76]

With sintering there occurs a decrease in surface area with a resultant decrease also in the surface free energy and the particle is densified in the bargain.

At least four steps are involved: (1) evaporation and condensation of atoms, (2) surface diffusion of atoms, (3) volume diffusion of atoms through the crystal lattice, and (4) bulk flow deformation.

The properties of solids important for sintering are:

Particle size and surface area (metals and ceramics must be powdered first then compacted, a process in which surface area is important).

Hardness (powdering is more difficult in harder units).

Density (the property looked for in sintering).

Surface forces (in sintering surface forces are lessened and the denser particle is more stable).

Cohesiveness (cohesion-but not too much-is important for compaction).

Particle size distribution (the strength of sintered metals depends on the particle-size distribution of the powder used in process).

I. DRYING [77,78]

Particle properties that are critical for efficient drying are:

Particle size and surface area (the amount of surface available for mass transfer of heat).

Porosity (nonporous particles easier to dry; practically no equilibrium moisture content).

Cohesion (very cohesive materials like $CaCO_3$ do not have a linear drying rate).

Melting point (low-mp solids must have a controlled lower temperature of drying).

Hygroscopicity, equilibrium, moisture content (these factors determine the amount of moisture to be transferred out and also the drying rate.)

Hardness (hard materials generally have a higher mp).

J. ADDITIVES AND COATING

These two operations are considered together because they have certain similarities. Additives and coating are usually applied in low amounts. Properties of solids important to both operations are:

Size-surface area: Small-size particles with their relatively greater surface area are very important for their covering or coating ability and also perform more efficiently as additives.

Surface forces cohesion and adhesion: These surface forces attract and hold a coating to larger particles.

Hardness: Harder materials have few attractive forces and are less suitable as additives or coating.

Flowability: Additives and coating materials usually are fed to the process. For this, their flowable nature is most important when designing the system. Flowability, of course, includes many other flow properties.

Hygroscopicity: Actually, a potential part of a material's flowable nature. Also, very hygroscopic coating particles cause problems.

Melting point: Materials with low melting points could also cause difficulties.

K. STORAGE [14,28,39,44,79,80]

Practically all of a material's characteristics are important in order to design a storage facility. The main properties that are critical to some degree are

Size-surface area: Finer particles have more of a tendency to mass.

Surface activity: Accelerates massing.

Hardness: Harder particles have less of a tendency to mass.

Hygroscopicity: Many problems related to massed materials in storage bins, or hoppers, occur because of the material's hygroscopic nature. Very hygroscopic materials need an inert gas atmosphere and specially sealed hopper cover. Constant humidity and temperature also help.

Melting point: Low melting-point substances can really cause a massed mess in a bin. Too large a mass must be avoided and temperature must be kept cool.

Flowability, cohesion, angles of flow, compressibility: A non-flowable material can really be a massed problem in a storage bin. The engineer must design accordingly.

Bulk density: The heavier the material, the greater the strength possible. The use of flow-conditioning additives also helps to reduce the massing tendencies of a problem material.

L. CLASSIFICATION AND PARTICLE SIZING [5,9,81,82]

Naturally, the most important property is particle size. Other properties that are critical are:

Surface and electrostatic forces: Adhesion of particles to sieve openings causes errors. May cause problem also in centrifuging, sedimentation, permeability, and adsorption methods.

Density: Important in sieving, sedimentation, and centrifuging.

Rugosity and shape: Important in sedimentation, permeability, adsorption, and filtration.

Hygroscopicity: Important in sieving, sedimentation, permeability, and adsorption.

Cohesiveness: Important in microscopy, sedimentation, permeability, adsorption, centrifuging, and filtration.

Among methods available for classifying solids according to particle size are (1) gravitational (elutriation), inertial, and centrifugal air classifiers, and (2) vibrating, oscillating, reciprocating, and centrifugal, static screens.

Table 19 lists 160 typical process materials along with their averaged properties, both physical and chemical. Along with the use of the caution code (see Table 20), Table 17 gives much pertinent and important information about each of the 160 materials. There are, in some cases, other grades of these materials which could be granular or powdered, or free-flowing or non-free-flowing. The material at hand must be material thoroughly investigated if there is any question.

The purpose of these two tables is to caution the engineer about each material listed. The caution code letters and numbers listed could possibly have errors due to transmission or printing. However, the use of this code should, in effect, lead to a thorough investigation of a material, especially if the coded information doesn't seem correct.

M. SUMMATION

From the foregoing review of the application of particle properties or characteristics to the different unit operations, the process engineer can realize the importance of a thorough knowledge of a gas-solid material. The most important and critical properties are: (1) size-surface area, (2) hardness, (3) bulk density and density, and (4) hygroscopicity (potential factor).

All the other important characteristics of a material are related to these four basic properties. Other than these inherent properties, the most imporatnt factors that lead to a material's possible process difficulties are time, exposure, and type of environment.

TABLE 19

Averaged Properties and Characteristics of Materials

Material	Averaged bulk density (lb/cu ft)	Possible problems[a]					Classification (Table 15)	Caution code (see Table 20)
		Flow	Dust	Abrasiveness	Hygroscopicity	Corrosiveness		
Alkalis	55	(√̌)	(√)		√	(√)	-	(G)MP(T)(1)(2)wx
Alumina, calcined	53						IA1	RTadijkn
Alumina ore	60			√			IA1	BRTadgiknp
Alumina, powder	20		√	√			IC1c	GIJQTahklnx
Aluminum, flake	28			√			IIC1	UX7igw
Aluminum, granules	65						IA1	UXigw
Aluminum, powder	50		√				IC1c	GIUXhgltx
Aluminum, sulfate, ground	64		√		(√)		IB3	BFMOST45adiw
Ammonium nitrate	60				√	√	IA2	DEGMOVXlabdfimoqsv
Ammonium sulfate	60				√	√	IA2	DEGMO3iops
Arizona road dust	61			√			IC3a	HIJPQTailn
Asbestos, long fibers	11	√					IIIB1	EFKLST9abegijklmnopqt
Asbestos, short fibers	21	√					IIIB2	EFKST9aefiklmpt
Bentonite, powdered	50		√				IC1a	GIJPhlx
Bentonite, granular	70						IA1	Pilw
Calcium carbonate, medium	48	√					ID1	EHKLPfiopq9
Calcium carbonate, heavy	60	√					IC3c	EHIKPfo9
Calcium chloride, anhydrous	55				√	(√)	IID2	GM17cegilmoprt
Calcium silicide	120						IB2	G(W)Zgikmt

a/ = yes, in some cases.

75

TABLE 19 (cont.)

Material	Averaged bulk density lb/cu ft	Possible problems[a]					Classification (Table 15)	Caution code (see Table 20)
		Flow	Dust	Abra- siveness	Hygro- scopicity	Corro- siveness		
Carbon, activated	20		✓			✓	IC1a	IGMOTZbhjq
Carbon, act., light	10		✓			✓	IC1c	IGMOTZbhiq
Carbon black	23		✓				IB6	GJQTZ9chpqt
Carbon black, light	15	✓	✓				IC2c	FIQTZ9chqt
Carbon black, granules	5		(✓)				IA5	GJLQTZ9chpqt
Caustic soda, flake	50				✓	✓	IID2	DNTVW1acegiklmprt
Caustic soda, pellet	73				✓	✓	IA1	DNTVW1acegiklmprt
Cement, Portland	76	✓	✓				IC3c	EHJKPSW59hoq
Chromite	118			✓			IC3a	BHJKQSTV9i
Chromium	158			✓			IB1	RTUVagijklnt
Citric acid	45	✓		✓		✓	IC1b	DFIM0128cehptuw
Clay, light	15	✓	✓				IC1a	O15h
Clay, medium	30	✓	✓				IC2c	F159h
Clay, heavy	70	✓					IC2b	BGJKQ5
Clay, lumps	75	(✓)	(✓)	(✓)			IA3	CFJKQ5(a)(d)(f)np
Coal	50	(✓)	(✓)	(✓)			IB1(2)	B(E)T(U)Za(e)ij(p)w
Coal, powder	30	✓	✓				IC2a	FIMT(U)Z5(f)x
Cocoa	30	✓	✓				IC3c	BDEHJKU589acefiopqtx
Copper	185						IB1	UXafinw
Copper sulfate	71					✓	IA2	GMDTV2iw
Corn, ground	32	✓					ID2	HKU58ceutix
Corn meal	41						IA5	GKU58ceimtuw
Corn sugar	38						IA5	DGKU458aceimtw
Dextrose	40						IB2	DGU458aceimtw

76

Material	Value							Code 1	Code 2
Diatomaceous earth, natural	8		✓	✓	(✓)			IIICS	HKLQT79hmq
Diatomaceous earth, calcium	9		✓	✓	(✓)			IC3c	HJKQT79hmq
Diatomaceous earth, flux calc.	16		✓	✓	(✓)			IIIC4	HJKQT79hm
Egg powder	21							ID2	DEHK589ceitu
Epsom salts	55							IIIC1	DK2fil
Ferric chloride	83	(✓)	✓	✓	(✓)		✓	IA1	NOT1acegijmnprtw
Flours	30		✓	✓	✓			IC2(3)c	EHJKU589cehmotu
Fluorspar	95		✓	✓	✓			IC2b	BHIKSTVix
Fluorspar, micronized	58		✓	✓	✓			IC2c	HIKSTVx
Flyash, light	35		✓	✓	✓			IC2b	IIIQ5T1
Flyash, sintered	43		✓	✓	✓			IA1	RSTaijnw
Flyash, heavy	68		✓	✓	✓	✓		IC1a	FIRSTacd1n
Gypsum	55		✓	✓	✓	✓		IC2b	(C)HK5
Hydrated lime, light	23	✓	✓	✓	✓	(✓)		IC3c	HJMPTX59bdfmoqv
Hydrated lime	35	✓	✓	✓	✓	(✓)		IC2c	HJMPTX59bdfmoqv
Iron	235+		✓	✓	✓			IB1	QUXafiov
Iron, powder	190		✓	✓	✓			IB2	GUXafiuv
Iron oxide, very light	18		✓	✓	(✓)	✓		ID1(C2c)	HK(L)Q9d(i)(p)
Iron oxide, light	40		✓	✓	(✓)	✓		ID2(C3c)	HK(L)Q9d(i)(p)
Iron oxide, medium	80		✓	✓	✓			IC3b	HKQ9dh
Iron oxide, heavy	160	(✓)	✓	✓	✓			IC3b	HKQ9dh
Lactose	44						–	IB1(6)	DGK458ceimotu
Lead, shot	420							IA1	TVXacegi
Lead oxide	175		✓	✓	✓			IC2c	HJKTVfi
Limestone, ground	95							IB2	GPQSfiw
Limestone, powder	63							IC2a	HIKPQShx

a/ = yes, in some cases.

TABLE 19 (cont.)

Material	Averaged bulk density lb/cu ft	Possible problems[a]					Classification (Table 15)	Caution code (see Table 20)
		Flow	Dust	Abra-siveness	Hygro-scopicity	Corro-siveness		
Magnesium	67						IA1	UWXYafivw
Magnesium carbonate, light	8.5	✓	✓				IC3c	HJKLP5c
Magnesium carbonate	15	✓	✓				IC2c	HIKP5
Magnesium oxide, light	22	✓	✓			(✓)	IC3a	HJKPST(W)X59dm(o)v
Magnesium oxide, medium	45	✓	✓			(✓)	IC2c	HIJPST(W)X59dm(o)v
Magnesium oxide, heavy	70	✓	✓			(✓)	IC2c	HIKPST(W)X59dm(o)v
Magnesite, dead burned	110			✓			IC2b	BCHJKPRSabefijlnpx
Malt, whole	34						IA1	DU58ceiuw
Manganese	95	✓	(✓)				IC2b	HJQTUVXagiv
Metals, light, granular	70			✓			IA1	Q(R)TU(X)agil(v)w
Metals, heavy, granular	190			✓			IA1	Q(R)TU(X)agil(v)w
Metal ores	170		(✓)	✓			(IB2)	GQ(R)T(U)abegijnqs
Mica, very thin	12	✓					IIA1	AFLSabegijklnpqt7
Mica, pulverized	40	(✓)	✓				IIA3	FITh
Mica, ore	33	✓					IIA1	(A)FST7abegijklnpqt
Milk, powder	38	(✓)	✓	✓			IC1c	DGIU48acehkmotu
Mineral ores, light	70	✓	(✓)	✓	(✓)		-	B(G)Q(R)T(U)(X)abegijln
Mineral ores, heavy	140	✓	(✓)	✓			-	B(G)QRT(U)(X)abegijlnps
Molybdenum	140	✓	✓				IC3b	FJKQ(U)afih
Molybdenum disulfide	79	✓					IC2a	FJKTadik
Oats, ground	27	(✓)					IB3	FKU578(c)egimotuw
Organic materials	40	(✓)	(✓)				-	DEG7(8)(c)(eg)ikmoqtuwx

Material					Code	Code2
Phenolic resin	30		√		IC1c	GITUabdghjlnqt
Phosphate rock	70		√		IC2b	HISSqx
Phosphate rock, lump	75		(√)		IB4	BCHKQS5a(b)egijmpqw
Pigments	34	(√)	√		ID4	EHK89egij(o)qt(L)
Pitch, granular	40	(√)	√		IB3	DGKQT(U)Z57abegimptw
Pitch, powder	21		√		IC2c	FIMQTZ5abhtx
Plaster of Paris	52		√		IC2a	EHJKW59h
Plastic powder	28		√		–	(A)FIUbdghktxj
Plastic pellets	35		√		IA1	(A)U(7)zlhwj
Potash	70	√	√	(√)	IC2b	EHIK45x
Pyrites	130	√	(√)		IB1	QSUX(6)agikw
Quicklime, pebble or lump	55	(√)		(√)	IA2(3)	(BC)GMPQSTWX5aegimopvw
Quicklime, ground	60	√	√	(√)	IB2	QMPQ3TWXabdgimovw
Quicklime, pulverized	55	√	√	(√)	IC3a	EHJKMPQSTWX5bfmovx
Quicklime, powder	42	√	√	(√)	IC2b	EHJKMPQSTWX5bcmovx
Salts, acid	45	(√)	(√)	(√)	–	(E)MNOT5a(e)io(q)twx
Salts, alkaline, powder	48	√	√	(√)	IC2	EFJKMPT5hx
Salts, alkaline, granular	55	(√)	(√)	(√)	IA1	MPT5iw
Saltcake	80	(√)			IC2a	(B)GJ45ix
Salt, rock	70			√	IA2	(BC)7aefiptN
Salt, table	60	√		√	IA1	N(5)(6)8acefitw
Salt, damp and wet	54	(√)	√	√	IA4	FKN568acegimoqtw
Sand, damp	71	(√)	√		IA1	RT(6)(a)dfi(1)(n)sw
Silica, very light	5	√	√		IA4	FRT6(a)dgi(1)(n)sw
Silica, light	15	√	√		IC2c	HJLRTcdfhlqt
Silica flour	50	√	√		IC2c	HJRTch
Soda ash, light	37	√	√	√	IC2c	HJKRT
Soda ash, dense	65			(√)	IC2b	DEHIKMPX35hx
				(√)	IA1	DMP3iw

a/ √ = yes, in some cases.

79

TABLE 19 (cont.)

Material	Averaged bulk density lb/cu ft	Possible problems[a]					Classification (Table 15)	Caution code (see Table 20)
		Flow	Dust	Abra-siveness	Hygro-scopicity	Corro-siveness		
Sodium bicarbonate	60						IC3b	DBEHJKPX5imovtx
Sodium nitrate	79					✓	IA1	DXY1acegijklmortv
Sodium tripolyphosphate	60		(✓)				IA1(B2)	(G)P5iw
Sodium silicate	44						IC3a	EHJK(N)PT45ax
Spices	35	✓	(✓)				IB3	FKU58icegitw
Starch, corn	35	✓	(✓)				IC1c	GIU5aehmlqtux
Starch, corn, pearl	44		(✓)				IB3	GU5eghmtuw
Starch, potato	45	✓	✓				IC2c	HIKU5beghmtux
Sugar	50				(✓)		IA1	DEU458abegikmtw
Sugar, XXX	30	✓	✓		✓		IC2c	DEHIU458abeghkmtx
Sulfur	35	✓	✓				IC2a	DEHJKMT9aceght
Sulfur, lumps	75	✓	(✓)				IB2(6)	D(C)EGKMT9acegit
Super phosphate	60	✓	(✓)				IB3	B(C)GOS5(e)ipw
Talc, light	12	✓	✓				IC3c	HIT5ac(d)hkm(s)
Talc, medium	28	✓	✓				IC2c	HIT5ac(d)hkm(s)
Talc, heavy	45	✓	✓				IC2c	HIT5ac(d)hkm(s)
Tin	227	✓					IC1b	GIUa(b)dghlnt
Titanium sponge	45			✓			IA1	BRUabdfijns
Titanium dioxide, light	31	✓	(✓)	(✓)			ID1	HLQ9(D)fikq(s)

80

Material	No.							Code 1	Code 2
Titanium dioxide	50		✓	✓			(✓)	IC3c	HJ(L)Q9(d)fhkq(s)
Trisodium phosphate, anhydrous	58				(✓)	(✓)	(✓)	IA1	MP3iw
Trisodium phosphate, anhydrous	65		(✓)		(✓)	(✓)	(✓)	IB3	MP3iw
Trisodium phosphate, hydrous	57						(✓)	IIIC1	DMP2iw
Tungsten	320			✓				IC2b	FJRT(U)abdgit
Tungsten carbide	480			✓				IA2	RTadegijt
Urea	46							IA1	DX4ikm
Urea lumps	47		(✓)					IA3	DCGX4ikmp
Wheat, ground	40							IID2	GKU58imouw
Wheat, grain	36							IA5	GKU58imouw
Wheat, whole	50							IA1	U58imouw
Wood flour	12		✓					IIIC3	MFUZ59adfimpt
Whey	35		✓					IC2b	DEHK48acegkptu
Zinc	192		✓					IC3b	GTUXgit
Zinc oxide	46		✓					ID2	EH(L)9adfioq
Zinc oxide, light	28		✓		✓			IC3c	HJK9ho
Zinc oxide, sintered	100				✓			IA2	GKQi1
Zinc, sintered ore	115							IA1	Rabdfijlps
Zinc stearate	8							IC3c	DEHJ(L)T9adhkmpq
Zircon, milled	115							IC3c	HJKRabdfs
Zircon	165							IA1	Radfis

$\underline{a}$/ = yes, in some cases.

TABLE 20

Caution Code

Particle characteristics

A. Electrostatic

B. Forms or can have hard lumps

C. Over 1½ in. lumps

D. Relatively low melting point

Flowability

E. Tendency to lump or mass

F. Will "hang up"

G. May "hang up"

H. Will arch

I. Floodable

J. Can be fluidized

K. Flow becomes poorer if packed

L. Special engineering to cause flow

Corrosion

M. Mildly corrosive

N. Highly corrosive

O. Acidic

P. Basic

Hygroscopicity

1. Deliquescent

2. Efflorescent

3. Cakes at room relative humidity

4. Cakes at over 81% relative humidity

5. Hygroscopicity affects flow

6. Damp

Special properties

7. Physical degradation

8. Contaminable

9. Packs under pressure

Unit operations

a. Size reduction difficult

b. Belt conveyor — difficulty

c. Belt conveyor — don't use

d. Screw conveyor — difficulty

e. Screw conveyor — don't use

f. Pneumatic conveyor — difficulty

g. Pneumatic conveyor — don't use

h. Fluidizable

TABLE 20 - Continued

Abrasiveness

Q. Mildly abrasive

R. Very abrasive

S. Can have abrasive impurities

Hazards

T. Dust harmful or irritant

U. Dust can be explosive with air

V. Toxic

W. Reactive with H_2O

X. Unstable — deteriorates oxidizes, etc.

Y. Fire hazard and/or oxidizing agent

Z. Flamable

Unit operations

i. Fluidizing difficult

j. Solid-solid blending difficult

k. Sintering — difficulty

l. Agglomeration — difficulty

m. Drying — difficulty

n. Coating — difficulty

p. Classification — difficulty

q. Dry feeder — difficulty or problem

r. Dry feeder — don't use

s. Bucket elevator — difficulty

t. Bucket elevator — don't use

Special properties

u. Microbial or biological degradation

v. Chemical degradation

w. Powder grades also

x. Granular grades also

REFERENCES

1. S. J. Gregg, Surface Chemistry of Solids, Reinhold, New York, 1951.

2. J. J. Bikerman, Surface Chemistry, 2nd. ed., Academic Press, New York, 1958.

3. A. W. Adamson, Physical Chemistry of Surfaces, 2nd ed., Wiley, 1967.

4. C. Orr, Jr., and J. M. Dallavalle, Fine Particle Measurement, Macmillan, New York, 1959.

5. R. D. Cadle, Particle Size, Reinhold, New York, 1965.

6. L. I. Osipow, Surface Chemistry, Reinhold, New York, 1962.

7. Robert F. Gould, Ed., Solid Surfaces and the Gas-Solid Interface No. 33, Advances in Chemistry Series, American Chem. Society, 1961.

8. J. J. Bikerman, Physical Surfaces, Academic Press, New York, 1970.

9. C. E. Lapple, "Particle-Size Analysis and Analyzers," Chem. Eng., Vol. 75, No. 10, May 20, 1968.

10. B. H. Kaye, "Determining the Characteristics of Fine Powders," Chem. Eng., Vol. 73, No. 21, Nov. 7, 1966.

11. A. Lieberman, "Fine Particle Technology in the Chemical Process Industries," Parts 1, 2, and 3; Chem. Eng., Vol. 74, No. 6, Mar. 27, 1967; Vol. 74, No. 7, Apr. 10, 1967; Vol. 74, No. 8 Apr. 24, 1967.

12. H. L. Bullock, "Six Tips on Processing Dry Particles," Chem. Eng., Vol. 74, No. 8, Apr. 24, 1967.

13. R. L. Carr, "Properties of Solids," Chem. Eng. Deskbook, Vol. 76, No. 22, Oct. 13, 1969.

14. R. L. Carr, "Particle Behavior, Storage, and Flow," ASME Publication 68-MH-6, Oct. 1968; or Brit. Chem. Eng., Vol. 15, No. 12, Dec. 1970.

15. R. Beach, "Preventing Static Electricity Fires," Chem. Eng., Vol. 71, No. 24, Dec. 31, 1964.

16. W. L. Lutes and H.F. Reid, "Selecting Wear Resistant Materials," Chem. Eng., Vol. 63, No. 6, June 1956.

17. National Safety Council, "Dust Explosions," Safe Practices Pamphlet No. 104, 1941.

18. M. Jacobson, J. Nagy, A. R. Cooper, F. J. Ball, and H. G. Dorsett, Jr., Bulletins 5733 (1961), 5971 (1962), 6516 (1964), 6597 (1965), "Explosibility of Agricultural Dusts," "Plastic Dusts," "Metal Powders," and "Carbonaceous Dusts," respectively, U. S. Dept. Interior, Bureau of Mines.

19. M. Jacobson and H. G. Dorsett, Jr., "Dust Can Be Dangerous," Modern Plastics, Vol. 39, No. 5, May 1962.

20. R. L. Carr, "Evaluating Flow Properties of Solids," Chem. Eng., Vol. 72, No. 2, Jan. 18, 1965.

21. R. L. Carr, "The Flow and Handling of Dry Materials Used in Water and Waste Treatment," Water and Sewage Works, Vol. 112, Nov. 30, 1965.

22. K. Terzaghi and R. B. Peck, Theoretical Soil Mechanics in Engineering Practice, Wiley, New York, 1948.

23. D. P. Koyninr, Soil Mechanics, 2nd. ed., McGraw-Hill, New York, 1947.

24. M. S. Ketchum, The Design of Walls, Bins, and Grain Elevators, 3rd. ed., McGraw-Hill, New York, 1919.

25. F. A. Zenz and D. F. Othmer, Fluidization and Fluid-Particle Systems, Reinhold, New York, 1960.

26. F. A. Zenz, "How Solid Catalysts Behave," Petroleum Refiner, Vol. 36, No. 4, Apr. 1957.

27. A. J. Stepanoff, Gravity Flow of Bulk Solids and Transport of Solids in Suspension, Wiley-Interscience, New York, 1969.

28. A. W. Jenike, "Storage and Flow of Solids," Bulletin 123, University of Utah, 1964.

29. D. C. H. Cheng, "Tensile Strength of a Powder," Chem. Eng. Sci., Vol. 17, p. 1405 (1968).

30. T. Tanaka, K. Gutoh, and K. Shinohara, "Cohesion of Fine Particles," Minerals Processing, Vol. 7, No. 3, Mar. 1966.

31. R. L. Carr, "Classifying Flow Properties of Solids," Chem. Eng., Vol. 72, No. 3, Feb. 1, 1965.

32. J. K. Rudd, "Bin Flow Simplified," Chem and Eng. News, Vol. 32, p. 344, Jan. 1954.

33. C. A. Lee, "Hoppers by Calculation," Chem. Eng., Vol. 61, No. 12, Dec. 1954.

34. C. A. Lee, "New Ideas About Hoppers," Chem. Eng., Vol. 59, No. 4, Apr. 1952.

35. C. A. Lee, "Hopper Design Up to Date," Chem. Eng., Vol. 70, No. 7, Apr. 1, 1963.

36. J. C. Smith, "Design a Hopper That Won't Arch," Chem. Eng., Vol. 62, No. 9, Sept., 1955.

37. J. W. Gossett, "Breaking Bin Bottlenecks," Chem. Processing, Vol. 28, No. 3, Mar. 1965.

38. R. M. LaForge, "How Research Points Way to Faster Hopper Flow," Material Handling Eng., Sept., 1965.

39. J. R. Johanson and H. Coligin, "New Design Criteria for Hoppers and Bins," Iron and Steel Engineer, Oct. 1964.

40. T. Tanaka, "Hopper Design and Material Flow," Minerals Processing, Vol. 7, No. 3, Mar. 1966.

41. K. Shinohara, "Designing Hoppers for Cohesive Material," Minerals Processing, Vol. 8, No. 4, Apr. 1967.

42. R. R. Towers, "Material Behavior-Key to Bulk Materials," Automation, Vol. 13, No. 8, Aug. 1966.

43. A. W. Jenike, "Better Design for Bulk Handling," Chem. Eng., Vol. 61, No. 12, Dec. 1954.

44. W. A. Gray, The Packing of Solid Particles, Chapman & Hall, Ltd., 1968.

45. J. Whetstone, "Solutions to Caking Problems of Ammonium Nitrate," Ind. & Eng. Chem., Vol. 44, No. 11, Nov. 1952.

46. W. L. McCabe, "Crystallization" (regarding caking), Perry's Chem. Eng., 3rd ed., 1950.

47. W. Scholl, "Reduce Caking Tendency," Chem. & Eng. News, Vol. 34, No. 36, Sept. 24, 1956.

48. R. R. Irani, C. F. Callis and Liu Ti, "Flow Conditioning and Anticaking Agents," Ind. and Eng. Chem., Vol. 51, No. 10, Oct. 1959.

49. R. R. Irani, H. L. Vandersall, and W. W. Morgenthaler, "Water Vapor Sorption in Flow Conditioning and Cake Inhibition." Ind. and Eng. Chem., Vol. 53, No. 2, Feb. 1961.

50. "Cationics Prevent Caking," Chem. Eng. News, Vol. 37, p. 52 (news item), Oct. 5, 1959.

51. F. C. Bond, "Control Particle Shape and Size," Chem Eng., Vol. 61, No. 8, Aug. 1954.

52. A. L. Stern, "A Guide to Crushing and Grinding Practice," Chem. Eng., Vol. 69, No. 23, Dec. 10, 1962.

53. F. J. Hiorns, "Advances in Comminution," Brit. Chem. Eng., Vol 15, No. 12, Dec. 1970.

54. A. Ratcliffe, "Crushing and Grinding," Chem. Eng., Vol. 79, No. 13, July 10, 1972.

55. C. W. Matthews, "Screening," Chem. Eng., Vol. 79, No. 13, July 10, 1972.

56. M. A. Buffington, "Mechanical Conveyors and Elevators," Chem Eng. Deskbook, Vol. 76, No. 22, Oct. 13, 1969.

57. W. G. Hudson "Why Use Pneumatic Conveyors?" Chem. Eng., Vol. 61, No. 4, April 1954.

58. J. Fischer, "Practical Pneumatic Conveyor Design," Chem. Eng., Vol. 65, No. 11, June 2, 1958.

59. M. N. Kraus, "Pneumatic Conveying," Chem. Eng., Vol. 72, No. 7, Apr. 12, 1965.

60. G. Broersma, "Pneumatic Transport of Fine Granular Material," ASME Paper, No. 68-MH-26.

61. D. Smith, "Pneumatic Transport and Its Hazards," Chem. Eng. Progress, Vol. 66, No. 9, Sept. 1970.

62. M. N. Kraus, "Pneumatic Conveyors," Chem. Eng. Deskbook, Vol. 76, No. 22, Oct. 13, 1969.

63. M. Sittig, "Fluidized Solids," Chem. Eng., Vol. 60, No. 5, May 1953.

64. J. Lowenstein, "Fluidization Velocities," (nomograph) Chem. Eng., Vol. 62, No. 4, Apr. 1955.

65. J. F. Frantz, "Design for Fluidization," Chem. Eng., Vol. 69, No. 18, Sept. 17, 1962.

66. T. R. Olive, "Solids Feeders," Chem. Eng., Vol. 59, No. 11, Nov. 1952.

67. P. A. Coffman, Jr., 'Guides for Selecting Chemical Feeders," Water and Sewage Works, Ref. and Data, Vol. 101, Nov. 1954.

68. A. F. Oszter, "Comments on Specific Feeder Applications," Can. Mining & Metallur. Bull., Mar. 1966.

69. C. A. Lee, "Solids Feeders," Chem. Eng., Vol. 71, No. 4, Feb. 17, 1964.

70. J. R. Johanson, "Feeding," Chem. Eng. Deskbook, Vol. 76, No. 22, Oct. 13, 1969.

71. S. S. Weidenbaum and C. F. Bonilla, "A Fundamental Study of the Mixing of Particulate Solids," Chem. Eng., Vol. 62, No. 1, Jan. 1955.

72. J. J. Fischer, "Solid-Solid Blending," Chem. Eng., Vol. 67, No. 15, Aug. 3, 1960.

73. C. Ludwig, "How and Why Solids Agglomerate," Chem. Eng., Vol. 61, No. 1, Jan. 1954.

74. J. E. Browning, "Agglomeration," Chem. Eng., Vol. 74, No. 23, Dec. 4, 1967.

75. "Selective Agglomeration," Chem. Eng., Vol. 76, No. 2, Jan. 27, 1969.

76. G. Dieter, "Powder Fabrication," Internat. Sci. Tech., Dec. 1962.

77. G. T. Tsao and T. D. Wheelock, "Drying Theory and Calculations," Chem. Eng., Vol. 74, No. 12, June 19, 1967.

78. A. D. Holt, "Heating and Cooling of Solids," Chem. Eng., Vol. 74, No. 20, Oct. 23, 1967.

79. J. F. Van Denburg and W. C. Bauer, "Segregation of Particles in the Storage of Materials," Chem. Eng., Vol. 71, No. 16, Sept. 28, 1964.

80. W. B. Pietsch, "Adhesion and Agglomeration of Solids During Storage, Flow and Handling--A Survey," ASME, Paper No. 68-MH-21.

81. L. T. Work, "Trends in Particle Size Technology," Ind. and Eng. Chem., Vol. 55, No. 2, Feb. 1963.

82. G. V. Syvertsen, "Sub-micron Particle Classification," Materials Eng., Vol. 74, No. 12, Dec., 1971.

83. "Felvation Speeds Powder Fractionation," Chem. Eng. News, Vol. 45, No. 7, March 6, 1967.

84. "Spherical Agglomeration," Ind. Eng. Chem., Vol. 61, No. 1, Jan. 1969.

Chapter 3

PNEUMATIC CONVEYING AND TRANSPORTING

C. Y. Wen

Department of Chemical Engineering
West Virginia University
Morgantown, West Virginia

and

William S. O'Brien
Department of Thermal and Environmental Engineering
Southern Illinois University at Carbondale
Carbondale, Illinois

1. CONVEYING SYSTEMS

A. HISTORICAL DEVELOPMENT

The movement of bulk solids from one location in an industrial process operation to another has often, in the past, been a source of considerable worry and expense, both in the design, construction and later in the operation of the equipment. While gases and liquids can flow rather easily from one location to another if given enough momentum or "push," granular solids are much harder to transport because they lack the internal ability to easily overcome inertia and move fluidly from one point to another.

Man has always recognized the ability of winds to move solid matter, as evidenced by the devastation wreaked by tornadoes and hurricanes. For centuries, man has utilized this power of the wind to move sailing ships and to turn windmills, thus transferring the momentum contained in the velocity of the moving air to the forward movement of the ship or the rotational movement of the windmill arms.

In the early 1900's, crude systems were commercially designed to move solids by lifting and carrying granules with a high-velocity air stream, an operation simulating a contained dust storm. These early pneumatic conveying systems were crudely designed and quite inefficient by today's standards; however, in most cases they successfully fullfilled the local industrial need for which they were designed.

Over the last half century, the art of designing pneumatic conveying systems has reached a rather high level of sophistication and acceptance, primarily through the successful attempts to utilize the many advantages of pneumatic conveyance in a multitude of solids-moving situations. The invention, development, and improvement of specific equipment items useful to the pneumatic conveying system, such as the Fuller-Kinyon fluid-solid pump (trademark of the Fuller Company, Catasanqua, Pa.) by Alonzo G. Kinyon from about 1910-20, helped stimulate the use and broaden the field of application of pneumatic conveyors. After the Second World War, the ease and flexibility that the pneumatic conveying system showed in moving the extremely large quantities of powdered cement required in major construction projects (such as the large hydroelectric dams in the western United States, the locks and pathways of the St. Lawrence Seaway, and numerous other industrial ventures) along with its easy adaptation for cleanly and safely loading and unloading grains, ores, coal, plastic pellets, catalysts, clay, cement, metal powder, alumina, and many other solid chemicals from ships, railroad tank cars, and hopper tank trucks, pushed gas-solid transportation to the forefront of methods used in moving solid particulate materials.

B. GENERAL DESCRIPTION OF CONVEYING SYSTEMS

The purpose of a pneumatic conveyor is to move solids from one location to another; i.e., from a tank truck to a storage bin, from a warehouse hopper to a reactor, etc. There are two major classifications of pneumatic conveyors, vacuum and pressurized, the designation depending on whether the pressure within the conveyor is more or less than the outside pressure. In general, the gas mover in the vacuum system is located behind the final solids collection device, while with the pressurized conveyor, the air blower is located in front of the equipment feeding the solids into the gas stream. The advantages and applications of these two conveying operations and the criteria for the use of each are discussed later in this chapter.

The major equipment items, in addition to the gas mover, are the solids feeder, the conveying conduit system, and the solids collection equipment. The feeding system must uniformly and continuously inject the solid particles into the gas stream. The conveyor lines must be designed so that the passage of the solid-gas mixtures is relatively free of any flow-restriction regions where the solids might settle and stop moving, causing partial or complete flow plugs and blockages. The collection device must separate the solids from the gas medium as completely and efficiently as possible. Industrial hygiene and air pollution control regulations dictate that this collection must be more than 99% efficient if the carrier air is to be discharged to the surroundings. The details and considerations involved in selecting and operating each of these equipment items are also discussed later in this chapter.

C. ADVANTAGES AND DISADVANTAGES OF CONVEYING SOLIDS PNEUMATICALLY

In discussing the advantage of using air or another gas to transport solid particles, one must naturally compare this method with other solids-moving systems, such as belt and screw conveyors.

The continuous operation of an industrial plant is often significantly more advantageous than a stagewise batch operation and, because of its greater reactive surface area per unit mass, the solid form most favored for further processing is the small-sized granular or powder shape. Of all the solids-moving equipment alternatives, pneumatic conveying is probably the most suitable method for the continuous transporting of small-sized solids. This same system is also readily adaptable for the large-volume batch movements of particles, i.e., from tank truck to storage bin.

A major advantage that pneumatic conveying holds over other systems is its extreme flexibility with regard to space design. The belt and screw conveyors are substantially one-directional, with a transfer point and a second independent conveyor needed if the path of the solids movement substantially changes its direction or elevation. As long as the conveying pipe is properly designed with long-radius bends and no major flow restrictions, the pneumatic conveyor can move the solids over, under, and around buildings, large equipment pieces, and other obstructions, keeping the conveying pipeline high to free valuable floor space for other operations. With the use of flexible hose and quick-connect fittings, the inlet and outlet ports can be manually positioned and connected with ease into the conveying system, thereby permitting the batchwise unloading or loading of portable hoppers, such as those on trucks, railroad cars, and barges. Since the various pumps, flow dividers, and receivers are very similar in operation to fluid flow equipment, most pneumatic

conveying systems can be readily automated and operated from a central
control station, a procedure that saves on the soaring costs of operating labor.

In comparison to other solids-moving methods, pneumatic systems
have a very decided advantage with regard to safety. Mechanical solids transporting devices, such as belts, screws, or buckets, offer
much potential for accidents, and elaborate precautions must be taken
for their safe operation. In the case of a specific cement plant
situation, a pneumatic solids transport system was reported to have
eliminated the source of 52% of all lost-time accidents attributed
to the screw conveyors and bucket conveyors, along with the drive
accessories, that the pneumatic system replaced [1]. In addition to
greater safety from mechanical injuries, a properly designed pneumatic conveyor reduces risk from fire and explosion hazards. A study
of the fire hazards involved in the high-velocity conveying of cellulose acetate dust in air demonstrates that the large degree of air
turbulence and the high velocities altered the combustion characteristics of the system so much that the explosion and combustion rates
were retarded, and no localized explosion could propagate throughout
the rest of the conveyor system [1]. The use of totally pneumatic
conveying systems has, in the past, led to a substantial savings in
insurance costs because of the lessened injury and fire hazards potential.

A properly designed and operated pneumatic conveyor also offers
a bonus in cleanliness and freedom from contamination. In the case
of vacuum conveying systems, any air leaks would be inward, and both
vacuum and pressurized equipment can be rather easily designed as
totally enclosed and sealed units. Thus, product contamination can
be held to a minimum. The major dust control problem points would
be at the inlet feeder and at the solids collection location; however,
careful design can render these locations virtually dustfree.

In the past, the major disadvantage of pneumatic conveying has
been the relatively high cost of power consumption compared to other
bulk-solids moving systems; however, modern-day equipment improve-

ments brought about by better understanding of the solids-in-gas flow
mechanisms have reduced these power costs considerably. These slightly higher power costs of pneumatic conveying are counterbalanced by
the steeply rising labor costs encountered with the other more manually operated solids-moving methods.

The type of material to be conveyed certainly has a very great
bearing on whether a gas-blown system can be utilized. While pneumatic conveying can be used for many solids-conveying applications,
these solids must be relatively dry and somewhat free flowing.
Properties of solid granules and powder that are important for pneumatic conveying have already been discussed in Chapter 2. As can be
seen from the discussion to be presented on the energy requirements
of the gas-solid transport, physical properties of the solids, such
as bulk density, size, and size distribution of particles; hardness
of solids; angle of repose; abrasiveness; explosion potential, are
critical in the design of pneumatic conveyance systems.

In general, solid matter that is fragile and easily crumbled is
not amenable to pneumatic conveying if that breakage cannot be tolerated in the final product. Hygroscopic or agglomerating materials
are not easily air-transported except in specially designed equipment.
Of course, easily oxidized solids cannot be conveyed by air; however,
an inert gas conveying system can be designed with a gas return line
for recycle purposes. Solid particles that are quite abrasive during
flow require special care in the selection of the metal used in con-
structing the conveyor and accessory equipment items. There is no
universally applicable design of a conveyor to handle all types of
solids. With difficult-to-handle solids, the knowledge of the art of
conveyor design and operation as developed through the years of application experience gives the major pneumatic conveyor equipment manu-
facturers a decided advantage over novice design engineers.

To date, pneumatic conveying is applicable for relatively short
distances, usually less than 10,000 ft. The major handicap in designing longer conveying lines is the difficulties encountered in design-

ing an operable in-line booster station. A breakthrough in developing such a gas-solid pump could permit the solids to be air-transported over practically any distance.

II. THEORY OF FLOW MECHANISMS

A. SOLID-GAS FLOW PATTERNS IN HORIZONTAL PIPES

Visual observation of the motion of solids in a glass pipe reveals that the flow patterns are rather complex and are affected, among many factors, by the solid-gas ratio, the Reynolds number of the flow, and the specific properties of the solids. As shown in Fig. 1, at very low solid-gas ratios, the solid particles are quite uniformly distributed throughout the pipe. This region, sometimes called the "homogeneous flow region," is characterized by the flow behavior in which the radial and axial solids density variations are so small that clusters of solids cannot be identified. There is evidence, however, that some of the particles are bouncing against the pipe wall during their travel [2]. As the solid-gas ratio is increased, the individual particles tend to settle at the bottom of the pipe and slide over other particles. When the particle segregation reaches a certain limit, the solids move from one dune to the next, under going alternate deceleration and acceleration. Additional increases in the solid-gas ratio can result in a slug flow which is characterized by the intermittent flow of gas and solids in alternating slugs. A higher solid-gas ratio may cause the solids to fill up a considerable portion of the pipe cross-sectional space. In such a high loading situation, the gas and the solid particles flow in the form of ripples, with the bulk of the solids remaining stationary. The solids loading in the ripple flow region may reach as high as 900 lbs of solids per pound of gas. The flows at a comparatively high solid-gas ratio are usually quite unstable, with large pressure-drop fluctuations.

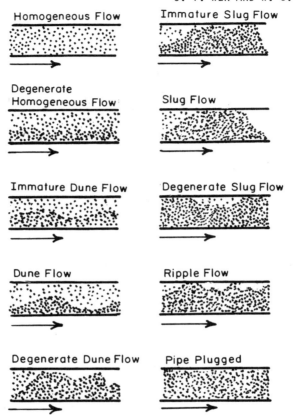

Fig. 1. Visual observations of the various flow patterns.

The correlations of pressure drop proposed by the various investigators, therefore, must be used with caution for the limited region of the flow condition investigated, and any extension of the relationships to another flow-pattern regime without a careful analysis of the actual flow condition encountered must be avoided.

This point can be illustrated more clearly from a schematic representation of the changes in the pressure drop per unit length of horizontal pipe, as shown in Fig. 2, as the superficial gas velocity is varied. The parameter W represents the fixed solids mass velocity.

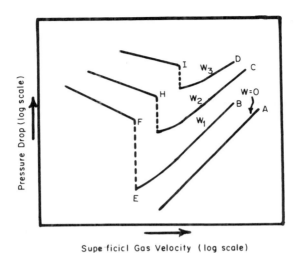

Fig. 2.   Variation of pressure drop in horizontally conveyed suspensions.

The line A represents the pressure drop for the gas alone flowing through a pipe. As the solid loadings, W's, are progressively increased, the pressure drops increase as shown by curves B, C, and D. If gas velocity is decreased, the pressure drop on curve B falls to point E, where the solid particles begin to deposit in the pipeline. As the solids deposit, they occupy an increasing proportion of the cross-sectional area of the pipe, thereby reducing the area of gas flow and causing the gas-phase pressure drop to rise to point F. At this point, the deposit rate equals the pick-up rate of solids due to higher gas velocities in the restrictive flow area of the pipe cross section. Further reduction of the gas flow rate causes an increase in pressure drop due to the additional solid deposition. The points F, H, and I correspond to solid-gas ratios which are the saturation carrying capacities of the gas stream; and the gas velocity at this point, called the "saltation velocity," is a function of the solid-gas ratio or the solids loading of the gas.

B. DILUTE-PHASE FLOW SYSTEMS

1. Horizontal Transport of Dilute Suspensions of Solid-Gas Mixtures

In the design of pneumatic transport systems, it is most important that the flow regimes in which the operation is to take place are identified.

The flow patterns of solid-gas mixtures differ considerably depending on the flow regimes as described in the previous section. In this section the simplest flow regime, that of homogeneous flow, is discussed in more detail. The homogeneous flow is defined as a flow in which the variations of solids density along axial and radial directions are small enough that clusters of solids and any settling of particles in the bottom of the pipe cannot be identified. It is clear, therefore, that in order to suspend the particles uniformly, the turbulent fluctuations must offset the gravitational effects. This suggests that homogeneous flow is likely to take place at a higher Reynolds number. A number of investigators have measured experimentally the time-averaged, radial concentration distribution of the solids [3 - 6]. These investigations can be summarized to provide the following rule-of-thumb for this type flow: If the Reynolds number is greater than about 10^5, the solid-gas ratio is less than about 2.0, and the particle size is less than about 200 μm, then homogeneous flow is realized.

The frictional energy dissipation for gas-solid flow is perhaps the most important in the engineering design and analysis of pneumatic systems. This problem has been reviewed extensively by many investigators [7 - 9].

The pressure drop in a dilute pneumatic transport system is typically as shown in Fig. 3 for the transport of wheat grains [10] and in Fig. 4 for the transport of coal powder. The total pressure drop ΔP_t is attributed to the sum of the pressure drop due to acceleration ΔP_a, the pressure drop due to the drag force on the particles ΔP_d, and the pressure drop of the gas caused by friction along the pipe wall ΔP_f. Thus,

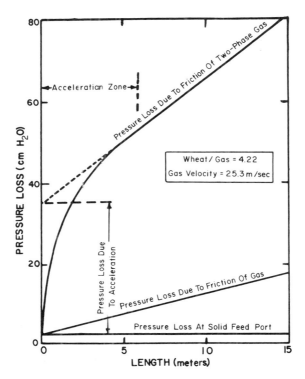

Fig. 3. Pressure loss in dilute-phase pneumatic transport,
Ikemori, [10].

$$\Delta P_t = \Delta P_a + \Delta P_d + \Delta P_f \cdot \tag{1}$$

The gas and particles are accelerated from the inlet until steady velocity conditions prevail a certain distance downstream. Approximately
6 m are required for this acceleration for the situation illustrated
in Fig. 3. The acceleration distance depends largely on particle
size (smaller particles require shorter acceleration distance) and
on the solids flow rate. It is not significantly affected by the gas
velocity or by the solid-gas ratio [11,12].

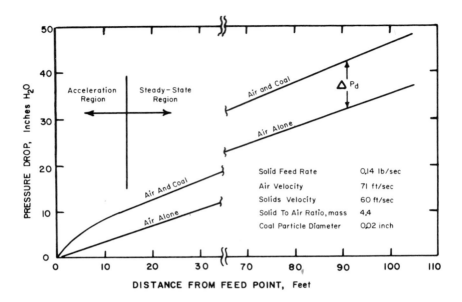

Fig. 4.   Pressure drop of 1-in. pipe carrying coal powder indicating particle acceleration region and pressure drop of air alone.

An analysis by Rose and Duckworth [12] of experimental results
for the transport of various materials indicates that, for pipes of
various angles of inclination upward from the horizontal position to
the vertical position, the acceleration length L_A is given by

$$\left(\frac{L_A}{D}\right) = 6\left[\left(\frac{G_s}{\rho_f g^{1/2} D^{5/2}}\right)\left(\frac{D}{d_p}\right)^{1/2}\left(\frac{\rho_s}{\rho_f}\right)^{1/2}\right]^{1/3} \tag{2}$$

2.   Pressure Drop and Forces Acting on Particles in Dilute-Phase Transport

a.   Drag Force.   The study of flow of gas-solid suspensions must
start with an analysis of the forces acting on a single particle.
The drag force acting on a single particle in an infinite span of flu-
ids can be expressed in terms of the drag coefficient C_{ds} as

$$F_{ds} = C_{ds}(\pi d_p^2/4)\rho_f(u_f - u_p)^2/(2g_c) \tag{3}$$

where

$$C_{ds} = 24\ N_{Re}^{-1} + 3.6\ N_{Re}^{-).313}\ \text{for}\ 2.0 < N_{Re} < 2,000,$$

and with

$$N_{Re} = d_p(u_f - u_p)\rho_f/\mu^f \qquad \text{(Wen and Yu [13]).}$$

This relationship of C_{ds} to N_{Re} is numerically presented in Table 1. presented in Table 1.

TABLE 1

Values of C_{ds} as a Function of N_{Re}

N_{Re}	C_{ds}
0.05	480
0.1	240
0.2	120
0.5	49.5
1.0	26.5
2	14.4
5	6.9
10	4.1
20	2.55
50	1.50
100	1.07
200	0.77
500	0.55
1,000	0.46
2,000	0.42
5,000	0.385

The drag force acting on one particle in a swarm of particles is, in general, larger than the drag force on a single particle flowing in an infinite span of fluid. This drag force on a particle in a multiparticle flow system can be related to the void fraction of the suspension as [14 - 15],

$$F_d = C_d (\pi d_p^{\,2}/4) \rho_f (u_f - u_p)^2 / (2g_c) \tag{4}$$

where

$$C_d = f(\varepsilon) C_{ds} \quad \text{and} \quad f(\varepsilon) \cong \varepsilon^{-4.7}.$$

Since in the dilute-phase pneumatic transport $\varepsilon = 0.8 \sim 1.0$, the drag coefficient C_d increases as particle concentration increases.

Let dW_s be the weight of solid particles in a horizontal pipe having a length dL. The number of particles, dn, in this section may be given by

$$dn = \frac{6 \, dW_s}{\rho_s \pi d_p^{\,3}}$$

Thus, if G_s is the solids flow rate in pounds per second, we have

$$dW_s = G_s dL/u_p \tag{5}$$

and

$$dn = \frac{6 G_s dL}{\rho_s \pi d_p^{\,3} u_p} \tag{6}$$

The void fraction ε in the section dL is then obtained by

$$\varepsilon = 1 - \frac{4 G_s}{\rho_s \pi D^2 u_p} \tag{7}$$

The total drag force acting on the particles in the section dL of

the pneumatic transport line is obtained by the summation of the drag
force on each individual particle. Hence, from Eqs. (4)-(6) we obtain

$$dF_d = \frac{3}{4} C_{ds} \frac{G_s \rho_f (u_f - u_p)^2 dL}{\rho_s d_p u_p g_c} \varepsilon^{-4.7} \tag{8}$$

The pressure drop caused by the drag force of the flowing gas on
the particles is given by

$$\frac{dp_d}{dL} = \frac{dF_d/dL}{\pi D^2/4} = \frac{3D_{ds} G_s \rho_f (u_f - u_p)^2}{\rho_s d_p u_p \pi D^2 g_c} \tag{9}$$

By letting $m = 4G_s/(\pi D^2 u_p \rho_f)$, the solid-gas ratio, Eq. (9) becomes

$$\frac{dP_d}{dL} = \frac{3}{4} C_{ds} \frac{\rho_f^2}{\rho_s} \frac{(u_f - u_p)^2}{d_p g_c} m \, \varepsilon^{-4.7} \tag{10}$$

Under steady-state operation the total pressure drop of the dilute
suspension flowing horizontally is the sum of the pressure drop due
to the drag force on the particles and the pressure drop of the gas
caused by the friction along the pipe wall, or

$$\Delta P_t = \Delta P_d + \Delta P_f \tag{11}$$

ΔP_f may be obtained from the conventional Fanning's friction factor
based on the corresponding flow rate of the gas phase without the
solid particles. Methods to calculate ΔP_f using Fanning's friction
factor can be found in textbooks dealing with fluid flows in pipes.

   b. Frictional Force on the Wall and Gravitational Forces. In
order to calculate the pressure drop of the flow of a solid-gas suspension in pipes from Eqs. (10) and (11), it is necessary to know the

particle velocity u_p. For a dilute suspension, the solid particle velocity can be obtained as follows [14 - 15]:

Let λ be the coefficient of friction due to the bouncing of the particles against the pipe wall and against each other. The frictional force of the particles can be represented by

$$dF_f = \pi DdL\tau_s = \frac{\lambda u_p^2}{D\ 2g_c}\ dW_s = \frac{\lambda u_p G_s}{D\ 2g_c}\ dL \qquad (12)$$

The values of the solid friction coefficients λ reported in the literature are listed in Table 2 for various types of particulate matter. Hinkle [17] obtained the following equation for the solid friction coefficient:

TABLE 2

Values of Solid Friction Coefficient

from Yang [16]

Substance	λ	Investigator
Tenite	0.004-0.008	Hinkle [17]
Polystyrene	0.008-0.019	"
Catalin	0.003-0.008	"
Alundum	0.009-0.018	"
Coal	0.005	Barth [18-19]
Coke	0.005	"
Wheat particles	0.003-0.013	"
Ottawa sand	0.010-0.021	Hariu and Molstad [20]
Sea sand	0.008-0.019	"
Microspheriodal cracking catalyst	0.008-0.023	"
Ground cracking catalyst	0.008-0.018	"
Rice	0.0058	Ikemori [10]
Soybean	0.0081	"

$$\lambda = \frac{2gD}{u_{ts}^2} \frac{u_f - u_p}{u_p} \tag{13}$$

where u_{ts} is the free-fall terminal velocity of a single particle.

Let u_t be the free-fall terminal velocity of an assemblage of particles having an average horizontal velocity u_p. The particles, while traveling a horizontal distance dL in the pipe, would fall a vertical distance $u_t(dL/u_p)$ Consequently, the gas must lift the falling particles against the gravitational force dF_g where

$$dF_g = u_t \; dW_s \; g/(u_p g_c) \tag{14}$$

The free-fall terminal velocity of an assemblage of particles can be given by the equation

$$u_t = \sqrt{\frac{4}{3} gd_p \left(\frac{\rho_s - \rho_f}{\rho_f}\right) \frac{\epsilon^{4 \cdot 7}}{C_{dst}}} \tag{15}$$

where

$$C_{dst} = 24N_{Ret}^{-1} + 3.6N_{Ret}^{-0.313} \text{ and } N_{Ret} = d_p u_{ts} \rho_f/\mu$$

The term u_{ts} is the terminal free-fall velocity of a single particle in an infinite span of fluid and can be obtained from

$$u_{ts} = \sqrt{\frac{4}{3} gd_p \left(\frac{\rho_s - \rho_f}{\rho_f}\right) / C_{dst}} \tag{16}$$

The equation of motion for the particles in a section dL is given by

$$dF_d - dF_g - dF_f = \frac{dW_s}{g_c} \frac{du_p}{dt} \tag{17}$$

Under steady-state conditions, Eq. (17) becomes

$$dF_d = dF_g + dF_f \tag{18}$$

Substituting Eqs (8), (12), and (14) into Eq. (18), we have

$$\frac{3}{4}\left(\frac{C_{ds}}{\varepsilon^{4.7}}\right)\left(\frac{\rho_f}{\rho_s}\right)\frac{(u_f - u_p)^2}{d_p g} - \frac{\lambda}{2}\left(\frac{u_p^2}{Dg}\right) = \sqrt{\frac{4}{3}\left(\frac{gd_p}{u_p^2}\right)\left(\frac{\rho_s - \rho_f}{\rho_f}\right)\left(\frac{\varepsilon^{4.7}}{C_{dst}}\right)} \tag{19}$$

Equation (19) provides the relationship between the particle velocity and the gas velocity. Particle velocities computed using Eq.
(19) have been compared with the experimental data obtained by Richardson and McLeman [21] with very good agreement [15]. Equation
(19) may be used to compute particle velocity when data on λ are
available. Knowing the particle velocity, the pressure drop in pipes
can now be calculated from Eqs. (10) and (11). Comparison between
the measured pressure drop and the pressure drop value calculated
using the procedure described above showed good agreement [14 - 15].

3. Vertical Transport of Dilute Suspensions of Solid-Gas Mixtures

The discussion provided in the preceding section for horizontal
transport can be applied to vertical transport, with allowances for
the differing effect of the gravitational force. The gravitational
force can be expressed as

$$dF_g = \frac{g}{g_c} \cdot dW_s \tag{20}$$

The drag force dF_d and the solid friction loss dF_f are the same as
those values expressed in Eqs. (4) and (12), respectively, except
that the velocity components are now in the vertical direction. The
pressure drop for vertical transport can be calculated using Eqs.
(10) and (11) as long as the particle velocity along the vertical
direction, u_p, is known. Yang [16], following a similar reasoning,
expressed the particle vertical velocity as

$$u_p = u_f - \sqrt{\left(1 + \frac{\lambda}{D}\frac{u_p^2}{2g}\right)\frac{4}{3}\frac{(\rho_s - \rho_f)d_p g}{\rho_f C_{ds}}\varepsilon^{4.7}} \qquad (21)$$

Equation (21) was reported to predict the experimental values of more than one hundred data points to within ± 20% accuracy [16].

Khan and Pei [22] presented a correlation of the pressure drops from dilute solid-gas suspensions flowing in vertical pipes based on their own data as well as those of previous investigators [20, 23 - 27].

4. Drag Reduction in Dilute Gas-Solid Flows

As discussed in the previous section, in most experimental investigations of flows of gas-solid suspensions it has been found that the addition of solid particles always increases the frictional resistance to flow. However, a few studies [28 - 31] reported a decrease in the frictional resistance below that of the pure gas when moving small particles at low solids loading ratios. For example, Rossetti and Pfeffer [28] reported that at a loading ratio of 1.5 and a gas Reynolds number of 25,000, the friction-factor ratio in a vertical test section was as low as 0.27 when transporting 30-μm particles in an air stream. This corresponds to a reduction in drag of close to 75%. The drag reduction has been explained as being caused by an interaction of the particles with the turbulence in the vicinity of the pipe wall. Additional studies are needed, however, before the benefits of this phenomenon can be utilized in industrial applications.

C. DENSE-PHASE FLOW SYSTEMS

1. Saltation Velocity in Horizontal Transport

The equations developed in the previous sections are for dilute solid-gas suspensions in horizontal transport lines. However, as the solid-gas ratio is increased, or the gas flow rate in a given solids carrying line is decreased, the solids begin to settle out along the bottom of the horizontal pipeline, partially obstructing

the flow area. The minimum fluid velocity required to carry the
solids at a specific rate without allowing them to settle out, the
"saltation velocity," is of particular importance in the design of
solid-gas transport systems. As discussed in the previous section,
the flow characteristics of dilute-phase suspensions are quite different from those of the dense-phase solid-gas mixtures, and it is
most imperative that the proper flow regimes be identified before
any of the equations presented there are applied. Zenz [32] presented an empirical correlation of the saltation velocity for single
particles flowing in horizontal transport lines. His correlation is
replotted in Fig. 5. It can be noted from this figure that, for
large particles, the saltation velocity u_s decreases with a decrease
in particle size, but below some specific particle size the saltation velocity becomes independent of size and then tends to increase
with decreasing solids size. A saltation velocity versus particle
size plot shows a minimum velocity, thus indicating the existence
of a region in which the finer particles are more difficult to maintain suspended, and a higher velocity would be required in order to

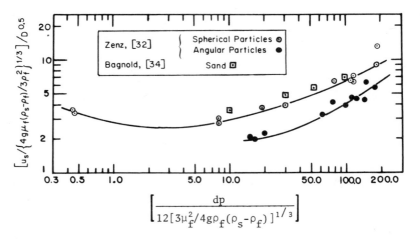

Fig. 5. Correlation of saltation velocity for single particle
size. Note: Terms in brackets are dimensionless, but D is in inches.

convey the particles in a dilute-phase suspension. This character is
attributed to the increased influence of surface effects and the
electrostatic charge effects in small particles [33].

To evaluate the saltation velocity for mixed sizes in a horizontal transport line, Zenz [32] suggests a method in which the saltation velocities for the smallest size and the largest size in the
mixture are first evaluated from Fig. 5. Then the slope S of the
line joining the two points as positioned in Fig. 5 is computed.
Finally the larger value of these two saltation velocities is selected, since it controls, and, with the slope S previously obtained,
Zenz suggests using the following approximate equation to calculate
the saltation velocity u_p' for the mixed-size particles:

$$W/\rho_s = 0.7(S)^{1.5}(u_p' - u_s)/u_s \qquad (22)$$

2. Pressure Drop in Dense-Phase Horizontal Transport

When the gas velocity decreases below the saltation velocity and
the solid-gas mass ratio becomes high, the particles begin to settle
and slide along the bottom of the pipe, as shown in Fig. 1. Such
dense-phase solid transport is usually achieved with fluidized bed
feeders.

A funnel-shaped delivery tube (upright or inverted) has been used
to deliver dense-phase solid-gas mixtures from fluidized beds. While
with conventional pneumatic conveying systems, the solid-gas loading
ratios are in the magnitude of 0.1-5.0, the transport lines using fluidized feeders may operate with solid-gas ratios of 25 to as high as
900. The modes of flow of these dense-phase solid-gas mixtures have
been discussed in the previous section. In the immediate vicinity
of the fluidized feeder, the solids appear to be still uniformly dispersed, but in the subsequent zones the solids tend to settle and
begin to form dunes at periodic intervals. Under such irregularities
in flow, the velocities of the particles would be expected to vary
considerably across the pipe cross section. In fact, the solid veloc-

ities seem to change from point to point along the pipe even after
the particles have left the supposed acceleration region, and they
undergo an alternating pattern of acceleration and deceleration. The
pressure drop should also be expected to fluctuate considerably.

   From the fact that the solids in a dense-phase transport move
more or less in collective bulk flow rather than individually, the
effect of individual properties such as size and shape may become
obscured [35]. Consequently, it is only logical to define the particle velocity as the average velocity of the mass of particles throughout the length of transport lines tested, a value computed using the
dispersed solids density. The dispersed solids density ρ_{ds} is defined as

$$\rho_{ds} = \frac{\text{weight of solids trapped within the test section by simultaneous closing of entrance and exit valves}}{\text{(volume of pipe used in the test section)}}$$

The average particle velocity u_p can be calculated from

$$u_p = \frac{W}{\rho_{ds}} \tag{23}$$

The dispersed solids density has been found in the dense-phase transport to be two to three times larger than the apparent solid-gas mixture density based on solid and gas mass flow rates [35], thus indicating a substantial slippage between solid flow and gas flow. Based
on such considerations, Wen and Simons [35] investigated the pressure
drop and solids velocity in three sizes of glass pipe carrying solids
from a fluidized feeder. The extent of slippage between the gas and
the solids is such that the gas velocities are roughly twice as large
as the solids velocities, with the slippage found to be practically
independent of the size of the particles in the range of the experiments conducted.

The pressure drop is correlated empirically by

$$\frac{\Delta P_t}{L\rho_{ds}} \left(\frac{D}{d_p}\right)^{1/4} = 2.5 v_p^{0.45} \tag{24}$$

The effect of particle shape and size on the pressure drop is rather small due to the fact that the solids travel predominantly along the bottom of the pipes as agglomerated masses.

In terms of gas velocity, Eq. (24) can be expressed as

$$\left(\frac{\Delta P_t}{LW_s}\right)\left(\frac{D}{d_p}\right)^{0.25} = 3.8(u_f)^{-0.55} \tag{25}$$

Equation (25) represents the pressure-drop data obtained by Holden [36] for coal particles using pipe diameters of 1, 2, and 3 in.; by Zenz [37] for glass beads, salt, and sand; by Wen and Simons [35] for coal and glass beads; by Carney [38] for glass beads; and by Koble [39] for coarse feldspar. The particle sizes covered range between 0.0028 and 0.066 in. Although Eq. (25) approximates the general trend of dense-phase pneumatic transport with solids-loading ratios ranging from 25 to 900, the correlation should be used with caution. Much uncertainty exists in determining the particle acceleration length and the effect of pipe diameter on pressure drop.

3. Choking Velocity in Vertical Pneumatic Transport

Vertical transport of solids is often encountered in catalyst regeneration units, such as in the operation of an air-lift tubular catalytic converter unit. For ease of understanding, the schematic representation of the flow characteristics in vertical transport as shown in Fig. 6 is helpful. At a fixed mass velocity of solids, W_1 or W_2, the pressure drop decreases as the gas flow rate is gradually reduced from point E or H to point D or G, respectively. Simulta-

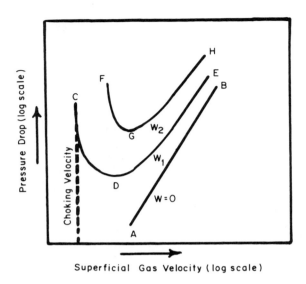

Fig. 6. Schematic representation of vertical pneumatic transport flow characteristics.

neously, the velocity of the particles and the voidage also decrease while the static head increases. At point D or G the solids head begins to exceed the gas friction and a further lowering in gas flow causes a sharp rise in pressure drop. As the point C or F is approached, the solids velocity has decreased so much that the gas can no longer support the high density of the suspension and the system collapses into a slugging flow pattern. The gas velocity at point C or F is called the "choking velocity" and represents the saturated carrying capacity of the gas stream. Therefore, the choking velocity is analogous to the saltation velocity in horizontal transport, since both represent saturation conditions for the solids carrying capacity between dilute-phase and dense-phase transports.

According to Zenz and Othmer [40], the choking velocity can be estimated from

$$W_{ch} = \rho_s (1 - \varepsilon_{ch})(u_{ch} - u_t) \tag{26}$$

where the critical voidage ε_{ch} is nearly independent of the density
of the solids. This critical voidage can be approximated by [40 -
41]:

$$\varepsilon_{ch} \simeq 0.03\rho_s + 0.91 \tag{27}$$

where ρ_s is in g/cm^3, $d_p > 0.17$ mm and $0.945 < \varepsilon_{ch} < 0.987$. For
uniform-sized particles, Zenz and Othmer [40] regard the choking
velocity as approximately the same as the saltation velocity, but
for mixed-sized solids, the saltation velocity is found to be three
to six times greater than the choking velocity.

Little data have been published describing the transport of solids as a dense phase where the solids density in the lift approaches
that in a static bed [42 - 43].

Sandy, Daubert, and Jones [43] present a large amount of data on
13 solid-gas systems. They point out that only the gas-solid friction and the work done by the gas to lift the solids need consideration in accounting for the total pressure drop in a dense-phase transport line. Analyzing their data, it is found that the observed
pressure drops are roughly four times those required to support the
solids in the lift lines. The average pressure drops in the lift
lines can therefore be approximated from the following equation:

$$\frac{\Delta P}{L} = 4\rho_b$$

where ρ_b is the bulk density of the solids and L is the length of
pipe. This approximate equation covers the following ranges of experimental data:

   particle density: 156-248 lb/cu ft
   particle diameter: 25-80 mesh
   solid-to-gas loading ratio: up to 5,000

In view of the difficulty in obtaining reliable data, an ample
safety factor must be included in designing the dense-phase vertical
transport system.

III. CONVEYOR DESIGN AND APPLICATIONS

A. CHOICE OF CONVEYOR SYSTEM

There are three general types of pneumatic conveyors, with several modifications and/or combinations of each type: (1) negative pressure conveying (vacuum), (2) positive pressure conveying, and (3) gravity movement of gas-buoyed particles. Each system offers distinct advantages for specific situations. Such advantages must be weighed carefully in making the final choice of the conveyor type to be used. Usually there is no clearcut choice among the types available and the final choice must be made by weighing the advantages and disadvantages of each versus the other possible alternatives.

1. Negative-Pressure (Vacuum) Systems

Vacuum conveying involves establishing conditions so that the solid particles are suspended or buoyed in the gaseous medium and then forced through a pipeline system by utilizing the energy of the gas expansions as the gas moves from the inlet points of greatest pressure (least vacuum) to the outlet port where the pressure is the smallest (most vacuum). The vacuum system offers two distinct application advantages. The feeding mechanism can be very simple and easily adaptable to situations where close dimension tolerances are required, such as the unloading of railroad box and hopper cars, hopper trucks, barges, and ship holds. Also, the system allows the conveying of material from any of several inlet feed locations to a single final outlet port location.

This conveyor system, illustrated in Fig. 7, is composed of a gas intake and a mechanical solids feeder (or a combined gas-solid intake), a solids transport line, a gas-solid separation device, followed lastly by the gas mover.

The solid particles are individually buoyed and drawn into the intake port by the inward movement of the air (such as the intake of a vacuum cleaner) or the solid particles are fed into the moving air

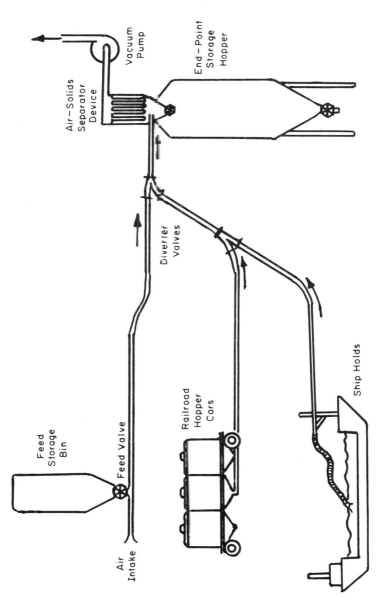

Fig. 7. Schematic of a negative-pressure conveying system.

stream by a mechanical feeder, such as a rotary star-valve or a two-valve lock-hopper feeder. The feeders, as well as the other individual component units in the pneumatic conveying systems, will be described and discussed later in this chapter. Because the flow-line pressures are less than atmospheric, any leakage in the lines would be inward, minimizing the dust problem along the conveying-line path. However, the lines should still be as leakfree as possible since undesired air inlet flow would require a larger air blower capacity than actually needed to move the solids. Also, inward air leaks could mean undesirable contamination of the solid product if the purity of the solid material is critical.

The terminal gas-solid separation device is usually located above the final single receiver bin, although the solids can be dropped into any of several closely spaced bins through diverter gates or a distributing screw conveyor. The separation of the solid matter from the gas may be in a dry cyclone, in the case of larger solid particles, or in a combination of dry cyclones and bag filter when a large proportion of the solid material is dust or fines.

After being cleaned of the solids, the conveying air stream finally enters the exhaust vacuum pump which provides the momentum for the flow system. This air mover is usually a rotary positive blower and is often followed by a suitable silencer to reduce the noise level.

Vacuum solids-moving systems have been designed for particle sizes as large as 2 in. in diameter and covering distances as long as 1,500 feet. The major uses of this vacuum operation take advantage of the system's quite simple solids-feeding techniques. Vacuums up to 12 in. of mercury and gas velocities from 65 to 135 ft/sec have been utilized in the design of these negative-pressure conveying systems.

2. Positive-Pressure Systems

The pneumatic transport of particulate solids in systems having pressures greater than atmospheric is used to move solids from a fixed feeding location to multiple delivery points using a single pipeline and diverter valves to direct the flow. Stoess [1] has classified the positive-pressure systems into three general types; the "low-pressure" system (limited to output pressures of about 20 psig), the "medium pressure" units requiring air pressures of from 15 to 45 psig, and the "high pressure" units, reaching from 45 psig to as much as 125 psig.

The positive-pressure conveyor equipment is usually arranged as shown in Fig. 8, with the feeder discharging the solids into a pressurized moving-gas stream. The gas stream can be diverted to any number of widely separated receivers by the use of pipe swithches, diverter gates, or gate valves. The major advantage of the pressurized system is the minimum amount of equipment required at the endpoint receiver, usually only a solids-particulate cleaning device on the air-stream vent leading from the final receiving hopper bin.

The low-pressure systems (0-12 psig) utilize rotary positive blowers to move the air stream and the medium pressure systems must use specially designed fluid-solid pumps, such as the Fuller-Kinyon pump. The high-pressure conveying systems (above 45 psig) usually employ a blow-tank feeder, with two tanks alternately and automatically filling, pressurizing, and blowing the solids into the transport lines with a rhythm much like reciprocating pistons.

The low-pressure system can handle dry pulverized, crushed, granular, and fibrous materials, while the medium and high-pressure systems normally can handle only relatively fine solids with a size range smaller than 50 to 100 mesh. The low-pressure systems are most adaptable for a large number of design applications, but the total conveying distances are limited to a few hundred feet. The high pressure systems are much less mobile and require greater care in design criteria selection, but they are well suited to long-distance conveying, with from 7,500 to 10,000 ft lengths feasible when easily fluidizable solids are involved.

Special precautions must be taken in the design and operation of the positive-pressure conveying systems to mimimize air-solids leak points and to prevent pressure buildup in the lines which cause slugg-

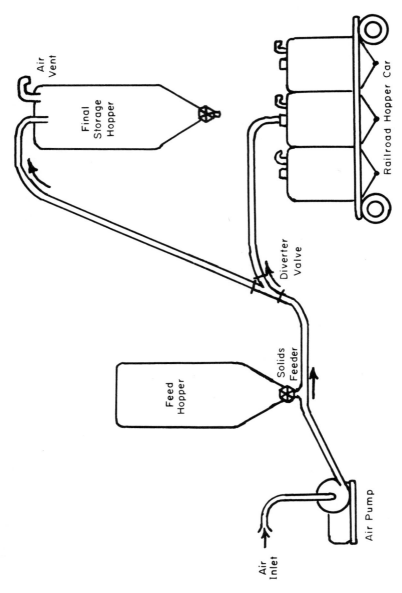

Fig. 8.   Schematic of a positive-pressure conveying system.

ing or a complete blockage of the solids flow. Of special importance
to the design engineer are the solids input feeder and the vent lines
on the receiving hopper. The solids feeder must distribute the solids
uniformly into the pressurized gas stream. A poor design of the feeding device could lead to pressure surges in the flow system backward
up into the feed hopper or it could lead to solids slugging or
plugs in the flow lines. Undersized or badly positioned vents in the
receiving hoppers can cause localized over-pressures leading to leaks
and flow stoppages. All lines, in-line valves, and piping bends must
offer a streamlined path to the gas-solid flow stream. The lines,
flanges, and joints must be leak-tight, but it is desirable that the
lines be designed for easy disassembly to save maintenance time if
and when line blockages do occur.

3. Special Combinations of Positive and Negative Pressure Systems

When the particular solids require an inert or special gas atmosphere to minimize the oxidation of the solids or inhibit any explosive
conditions, a closed-loop system must be designed to recycle the gas
stream. A duplicate return flow line is required, as pictured in
Fig. 9. The addition of such lines causes an increased capital construction cost, but a savings occurs in the quantity of inert gas
generation required, and also, 100% removal of the solids at the re-
ceiving-end product bin is not required since the solid dust and fines
can be continually recirculated. The dual lines do offer more problems of cleanup if the conveying system is used for the transport of
several different product materials, and the chances of cross-con-
tamination and purity losses are increased.

If the solids must be moved from several origin points to a multiple number of receiving vessels in widely scattered locations, a
combination vacuum-pressure system can be designed, as shown in Fig.
10. Because the designing of a pump or blower to move gas streams
containing high concentrations of solids is quite difficult, most of
the solids must be removed from the gas stream, the gas phase repres-
surized, and then the solids refed back into the pressurized stream

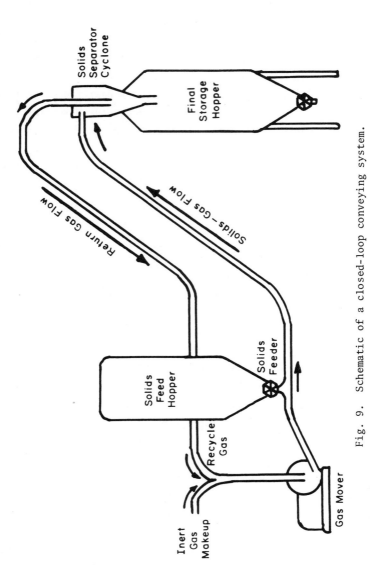

Fig. 9.   Schematic of a closed-loop conveying system.

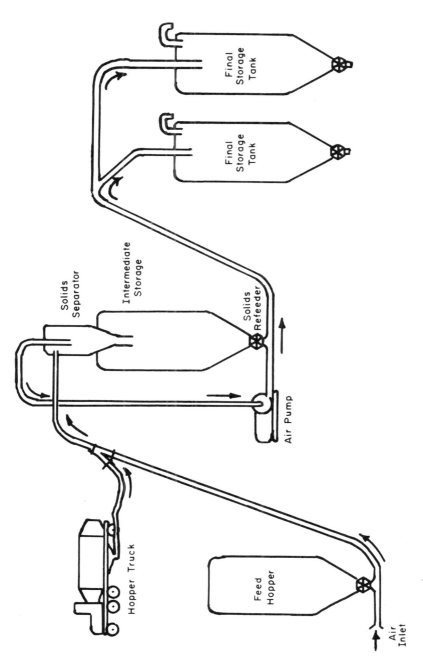

Fig. 10. Schematic of a negative-positive pressure conveying system.

of gas. While this design technique does offer much choice in feed
origin and terminal delivery locations, the solids separation and
the refluidizing operations around the gas pump definitely add an
area where operating problems could develop. This dual system offers
a design advantage in that the intermediate separator-blower-feeder
equipment can be located in a central permanent position and a minimum of equipment would be needed at the solids feeding location and
at the ultimate receiver location. A similar equipment arrangement
that separates the solids, repressurizes the air, and resuspends the
solids could also be employed as a "booster station" in a long-distance pneumatic conveyor, thereby practically doubling the effective
operating distance; however, the operating pressures of these long-
distance conveyors are usually quite high.

4. Gravity Movement of Gas-Buoyed Solids
     The conventional pneumatic conveying device requires the complete
buoyance of the solids into the air stream with the solids being
moved by the momentum motion of the air.
     A modification of this conception is offered by the air-activated
gravity conveyor, where the dry solids are dropped onto a fabric-
covered frame tilted at a slight angle, up to 15°, with air passing
upward through the fabric at a flow rate intermediate between the onset of fluidization and the terminal velocity of the solid particles.
Once the particles become fluidized or air-buoyed, they tend to move
by gravity in the direction of the slope of the conveyor frame. The
conveyor can be designed to change directions in long-radius curves,
and adjustments can be made to control the delivery flow rate, within a limited range, by varying the slope of the conveyor.
     With this type of conveyor transport distances and elevation
changes are somewhat limited; however, a much smaller quantity of
air is required at a relatively low pressure for this system than is
required for the conventional pneumatic transport of the solids.
With the larger gravity-movement conveyors, the system must be carefully designed to ensure a uniform and proper flow rate of the gas

stream through all portions of the fabric surface of the conveyor.
Of course, elevation changes are required and must be from the feeder
in the highest position down to the terminal endpoint at the lowest
elevation. By proper selection of the temperature and other properties of the fluidizing gas stream, the solids could be, for example,
dried, cooled, heated, or chemically treated during their transport
from a hopper to a reactor, thereby accomplishing two or more tasks
at the same time. This technique is also used to buoy and gravity-
move solids down the sloping inside walls of solids hoppers to insure
that the solid matter will smoothly move downward and not bridge over
to block the outlet port at the hopper bottom.

B. EQUIPMENT COMPONENTS

While an engineer inexperienced in pneumatic conveying could design new apparatus or adapt available equipment which could be used
in an air-solid conveying system, the myriad of special design considerations of the various conveying equipment components makes the
design of such a system for maximum efficiency and troublefree operation a subject where the state-of-the-art knowledge and prior experience are quite necessary. For the designing details, operating considerations, and a more complete understanding of this specialized
equipment field, the reader is urged to refer to the special literature provided by the pneumatic conveying equipment design companies,
to the several reviews [44-45], to the two major reference books in
this field [1,46], and to the proceedings of several symposiums
specifically dealing with this subject area [47-49]. Kraus [46] in
his book Pneumatic Conveying of Bulk Materials presents a very fine
discussion and description of the individual items of equipment which,
when combined, comprise a pneumatic conveying system.

1. Solids Feeding Devices

The most frequently used device for feeding the solids uniformly
into either the negative-pressure (vacuum) or the low-positive-pres-
sure conveying system is the rotary air-lock solids feeder or "star-

wheel" feeder. This device consists of a cylindrical chamber (with horizontal axis) in which a rotating paddle-wheel or star-wheel is positioned. The radial vanes of the paddle-wheel are of dimensions, such that the tips of the vanes are in air-tight contact with the walls of the feeder housing shell. The incoming solids continually fill the topmost compartment, with the star-wheel rotating to the exit bottom position where the solids drop or are gas-swept into the moving gas stream. By varying the rotation rate of the inner star-wheel, the solids feed rate can be metered with fairly accurate precision.

Rotary feeders must be specially designed to match the characteristics of the solid material being meter-fed into the gas stream. The particle size and abrasion characteristics of the solid matter have a decided influence on the proper design of the feeder vanes, giving particular design consideration to the tips of the radial vanes which must have a very close clearance (about 0.005 in.) with the walls of the cylindrical housing walls to provide an adequate gas-seal.

Precautions must also be taken in designing the feeder to eliminate any points in the solids flow path where localized pressure buildup might cause solids flow stoppage. For instance, if the valve is feeding solids into a positive pressure system, provision should be made to vent the pressurized gases from the "empty" valve compartments while the feeder star-wheel compartment is being rotated back up to the upper inlet solids feeding position; otherwise, the higher gas pressure in the empty compartment might block the solids from filling the entire compartment chamber. The compartment chamber must also be designed for complete filling and emptying, a special problem if the solids tend to be sticky and agglomerate easily. If the solids are hygroscopic, the hopper might have to be purged with an inert dry gas or the feeding system heated by steam trace lines to prevent localized condensation and solids buildup.

Since the rotary valve is often the source of many maintenance and plugging problems, its physical location for easy access to disassembly, repair, and/or replacement is most important.

While the rotary feeder is probably the most widely used solids feeding device, quite a number of other techniques are being used to successfully inject the solids into the flowing gas stream.

With the vacuum conveying system, a simple combined air-solids intake feeder can be used to great advantage in the manual or semi-automatic emptying of railroad boxcars, shiphold chambers, and open-bed trucks. The intake nozzle, usually at the end of a length of flexible tubing long enough to reach the furthest corners of the solids-containing chamber, is manually directed at the pile of solid matter and moved around to assure a continuous intake of the solids. The major advantage of this technique is its mobility and adaptability to different types and shapes of incoming solids containers (railroad cars, trucks, etc.). However, this system does not automate readily and does require manual operation and/or supervision at all times. Also, the intake solids flow tends to be at a nonuniform rate, and often the air-solids mass ratio is substantially greater than the most efficient proportion with respect to the power supply required (minimum air blower capacity needed to move the solids).

With a low-pressure conveying system, a solids feeding device designed similar to the water- or steam-fed gas vacuum siphon can be used to vacuum-pull the solids into the gas stream. In this device, the solids must be quite free flowing and easily fluidizable.

A medium-pressure solids-conveying system would mean a greater than one atmosphere pressure differential across the solids-feeder, too much of a differential to maintain the air-tight seal through the standard rotary-feed valves. It is with this system, used for the medium-distance range (200-2,000 ft) conveying of solids, that the specially designed fluid-solids pump was found to be most successful. Such a pump, invented by Alonzo G. Kinyon before World War I and for which he was awarded the Edward Longstreth Medal in 1926 by the Franklin Institute of Philadelphia, involves the feeding of the solids with a screw conveyor into a special nozzle-shaped forward chamber of the pump where the compressed air stream (15-45 psig) fluidizes the solid particles and provides the forward momentum for their pneumatic conveyance [1,50]. This type of fluid-solids pump is quite successful and relatively maintenance-free in moving fine-sized (200 mesh), easily-fluidizable solid particles.

The transportation of solids for relatively long distances, up to 2 mi., requires greater positive pressures than can be handled in a present-day fluid-solids pump. In these 45-125 psig systems it is better to feed the air into the solids rather than meter the solids into the moving air stream, as is done in the rotary-star feeder and in the medium-pressure fluid-solids pump. The air-into-material solids feeder usually involves a specially designed blow-tank [46], in which alternately and in sequence, the tank is charged with fresh feed solids, the tank is pressurized to semi-fluidize the particles nearest to the outlet port, and then the pressure is suddenly released through the solids discharge port to "blow" the solids into the transport line. Although the cyclic rhythm of operating the single blow-tank is pulsating, a nearly uniform solids feed flow can be maintained by utilizing a pair or more of blow tanks designed to automatically feed solids into the transport line in an alternating sequence.

2. Piping Lines, Gates, and Flowmeters

The conveying pipe can be constructed of the many materials available today, the material selected to match the properties of the solid matter being transported within the constraints of availability and cost. The path of the piping should be as straight as possible from inlet to delivery point, with as few direction changes as can be physically tolerated; however, path directions can be planned to pass around, over or under large obstructions such as interfering large equipment items.

All direction changes should be made using long-radius bends, with the bend-radius being at least six to twelve times the pipe diameter. Pipe joints should be butt-welded or be a compression-type sleeve coupling to eliminate any crevices or points of potential material-flow obstruction or blockage.

In order to direct the flow of gas and solids to one receiving tank among several choices, diverter in-line valves or gates are

used. Many types of diverter valve design are commercially available,
the main design criteria being a streamlined solid-gas flow path with
limited path obstructions and no sharp direction changes. Because the
valves and gates are often the location of solids plugging, the valve
should be designed and located for easy disassembly and replacement.
If the conveying system is rather simple, the path direction could be
changed (such as in selecting the particular solids receiving bin) by
the manual switching of a flexible connecting line using quick-dis-
connect line couplings. This saves the cost of the rather expensive
diverter valve and its control system, and eliminates any problems of
a faulty valve operation as well. Although the manual switching and
connection of the lines means more manual labor time and cost, there
would be a decreased chance of error in the selection of the desired
flow path direction by inattentive operators.

Other equipment components are commercially available that aid
in the transporting of certain specific solids. If the solids happen
to be tacky and tend to agglomerate and form larger particles or
lumps, in-line rotary chopping devices are available to reduce the
particle size back to a more fluidizable range. In-line mixers and
blenders are available to homogeneously blend two or more types of
solid matter during the transportation. Provision can also be made
to heat or cool the solids, dry them or, perhaps, chemically treat
them while they are being conveyed from one location to another.
The design of such a treatment device would follow the chemical kinetic relationships derived for a plug-flow reactor system.

Several in-line flowmeter devices have been designed to measure
the solids flow rate through the pipe lines, and these flowmeters
are rather successful, with a ± 5% accuracy. One device is the Fibre
Optics Probe developed by Peskin [51] in which the signal correlating
the flow rate of the solids is generated and attenuated by the blockage of a light beam by particles passing between two 200-μm diameter
fiber optic rods bent with the tips directed at each other, with the
distance between the ends of the fiber tips fixed at 0.011 in. Another device, developed by the U.S. Bureau of Mines at the Morgantown,

W. Va., Coal Research Center, for measuring powdered coal suspensions, involves a venturi throat in the gas-solids flow line, followed by a positioned strain-probe consisting of a circular target attached to a rod at the end of a cantilevered metal strip with attached foil-type strain gages [52-53]. A flowmeter of this type, called the Target Flowmeter was developed into commercial applications in 1966 by the Foxboro Company.

3. Solids Separation Devices

The methods used at the endpoint receiving bin to separate the solids from the conveying gases have received much attention and have been considerably improved during the recent campaigns to reduce air pollution and to suppress the dust problems. With positive-pressure conveying systems, the solids-laden gas can be directly charged into the terminal solids-receiving bins, and the exiting gases passing from the bins can be charged through the top vents into the surrounding atmosphere. However, this technique is satisfactory only if the solids are absolutely dust-free or if there is a properly designed dust filter installed in the vent lines. Provision can be made to clean the solids from the gases by fabric filters, but usually provision must be provided to periodically physically clean the fabric of accumulated clinging solids either by mechanically rapping or shaking the filter bags, or by reversed air pulses blown back through the fabric countercurrent to the normal air flow direction.

The selection and design of the solid-gas separation devices is as specialized as the design of pneumatic conveying-system apparatus. The selection and sizing of the filtering equipment is based mainly on past experience and on actual pilot plant scale testing that utilizes the specific type of woven fabrics under test conditions which simulate and duplicate, as nearly as possible, the operating conditions of the proposed full-sized plant.

Many of the solids separation devices involve both a mechanical separator, such as a large single dry cyclone or a set of mini-cyclones, which remove the bulk of the solids, followed by a fabric-

type dust collector which, with proper design, can remove the rest of
the solids. The combination of the dry cyclone and the bag-type dust
collector should remove well over 99% of the solids content from the
gas stream.

Particle size plays a great part in the design of the dry cyclone
separators. Particles as small as 10 μm in diameter can be separated
to within 99% removal efficiency in a properly designed high-effici-
ency dry cyclone. But this extra good separation of the solids
from the gas stream is made at the expense of rather large gas-phase
pressure drops, and a small degree of collection inefficiency of the
cyclones could be tolerated as long as the followup fabric filters
are operable.

The fabric filter devices can vary in design from relatively
small package filter units which can be purchased complete with auxiliary blowers and hopper bins for collected solids, to extremely
large custom-designed units. The solids-gas separators are designed
with special consideration given to the solid-gas loading of the conveying stream, the size and physical properties of both the solids
and of the conveying gas, and the thoroughness to which the gas
stream must be cleaned. Methods to clean the fabric filters during
the solid-gas flow (or between the flow periods if the flow is intermittent, such as during the periodic filling of a storage bin) can be
accomplished by reverse-air flow blasts and by shaking with or without
the air pulses. The particular fabric used in the filter bags must
be chemically inert to the solids and to the gas medium, and the system must be able to shake completely clean during the cleaning cycle.
If the dust particles happen to be electrostatically highly charged,
the use of a slight electrical charge on the collector fabric might
aid in attraction and removal of the particles from the gas stream,
while a reversal of this electrical charge might help to strip the
solids from the fabric surface during the cleaning step. The fabric
must be tough enough to withstand the repeated cleaning shake cycles
and to lose little of its physical strength with age, since bag replacement can be rather expensive and time consuming.

As with the selection of most of the pneumatic conveying equipment components, much consideration must be given to the availability and ease of maintenance and repairs. If the maintenance labor force in the company is limited in number or is not highly skilled, the various equipment devices must be designed to operate with as little trouble and repair time as possible.

C. INDUSTRIAL APPLICATIONS

The profitable application of pneumatic conveying has been made in a myriad of industrial situations and the reader is referred to the many construction and industrial trade journals for the descriptions of the most successful specific applications.

The use of pneumatic conveying systems offers definite advantages in the loading and unloading of railroad cars, hopper trucks, ship holds, and other shipping containers, as compared to other solids-moving techniques. This advantage is emphasized by the use of mobile containers which have been specifically designed for pneumatic loading and unloading, instead of the older railroad boxcars and highway trucks which were not designed to maximize the use of pneumatic solids transport. Practically any fluidizable solid particulate material could be unloaded and transferred to storage bins with little trouble. For convenience of operation in remote locations or in locations with limited equipment, some of the trucks or other bulk container vehicles are equipped with their own portable or mounted pneumatic conveying equipment components (air compressor, dust collector and a limited length of flexible conveying pipelines).

A very good description of the use of air-solids conveying techniques for moving various types of industrial solids is given by Stoess in Chapter 4 of his book Pneumatic Conveying [1]. The reader is referred to this text for the details of using pneumatic conveying in the industrial handling of cement; flour and other bakery ingredients; hops, malt, and other brewing industry compounds; kaolin; plastics and rubber pellets; pulp and paper industry solids; clays; feed mill grains (including corn, sorghum, oats, barley, and wheat); water purification chemicals; and many industrial inorganic salts.

Other recent articles describe the application of pneumatic transport in the movement of foodstuff [54], sugar [55], sawdust [56], cement [57], coal [58-60], garbage [61], water treatment chemicals [62], gypsum [63], and crushed rock [64].

Much application information has been presented over the past 20 years in public literature, but the commercial pneumatic conveying equipment firms have privately accumulated an even larger amount of the state-of-the-art application knowledge. Although pneumatic transport does offer great advantages over the other alternative bulk-solid conveying systems in many situations, there are some types of solids that cannot be moved in this way, and other solids that require special modifications in the standard conveying equipment design and/or operation. Consultation with persons experienced in this field is certainly urged before one attempts to design and construct a system to move large quantities of a solid about which little information has been published in literature.

If improperly designed and badly operated, a gas-solids tranportation system could be a constant headache and a financial waste. But a properly designed system could be quite inexpensive to operate and almost troublefree throughout its many years of operating life.

REFERENCES

1. H. A. Stoess, Jr., Pneumatic Conveying, Wiley-Interscience, New York, 1970.

2. O. Adam, Chemie Ing., 29, 151 (1957).

3. K. Goto, and K. Iinoya, Kagaku Kogaku, 28, 73 (1964).

4. R. L. Peskin and H. A. Dwyer, "A Study of the Mechanics of Turbulence Gas-Solid Shear Flow," paper presented at ASME Chicago Meeting, November 1965.

5. S. L. Soo, G. J. Trezek, R. C. Dimick, and G. F. Hohnstreiter, Ind. Eng. Chem.; Fundamentals, 3, 98 (1964).

6. D. G. Thomas, A.I.Ch.E. J., 10, 517 (1964).

7. R. G. Boothroyd, Flowing Gas-Solids Suspensions, Chapman and Hall Ltd., 1971.

8. S. L. Soo, Fluid Dynamics of Multiphase Systems, Blasidell, 1967.

9.   D. G. Thomas, A.I.Ch.E. J., 9, 310 (1963).

10. K. Ikemori, J. Mech. Eng. Japan, 62, 480 (1959).

11. L. Papai, Acta Tech. Hung., 14, 95 (1955).

12. H. E. Rose and R. A. Duckworth, "The Fluid Transport of
     Powdered Material In Pipelines," Proceedings of I. Chem. E.
     VIG/VDI Joint Meeting, 1968, p. 50968.

13. C. Y. Wen, and Y.H. Yu, Chemical Engineering Progress Symposium
     Series, 62, No. 62, 100 (1966).

14. C. Y. Wen, "Pneumatic Transportation of Solids," Proceedings of
     Institute of Gas Technology-U.S. Bureau of Mines, Morgantown,
     W. Va. Oct. 1965, Information Circular No. 8314, U. S. Bureau
     of Mines, 1966.

15. C. Y. Wen, Bulk Material Handling, Vol. 1, M. C. Hawk, Ed.,
     University of Pittsburgh, Pittsburgh, 1971, p. 258.

16. W. C. Yang, Ind. Eng. Chem.; Fundamentals, 12, 349 (1973).

17. B. L. Hinkle, Ph.D. Thesis, Georgia Institute of Technology,
     Atlanta, Georgia, 1953.

18. W. Barth, Chem. Ing. Tech. Z., 30, 171 (1960).

19. W. Barth, Chem. Ing. Tech. Z., 32, 164 (1962).

20. O. H. Hariu and M. C. Molstad, Ind. Eng. Chem., 41, 1148 (1949).

21. J. F. Richardson and M. McLeman, Transactions of the Institute
     of Chemical Engineers, 38, 257 (1960).

22. J. I. Khan and D. C. Pei, Ind. Eng. Chem.; Process Design and
     Development, 12, 428 (1973).

23. S. S. Chandock, Ph.D. Thesis, University of Waterloo, 1970.

24. L. Farbar, Ind. Eng. Chem., 21, 1184 (1949).

25. J. H. Jones and H. D. Allendorf, A.I.Ch.E.J., 13, No. 3, 608
     (1967).

26. K. V. S. Reddy, Ph.D. Thesis, Univ. of Waterloo, Waterloo, Can.,
     1967.

27. E. G. Vogt and R. R. White, Ind. Eng. Chem., 40, 1731 (1948).

28. S. L. Rossetti, and R. Pfeffer, A.I.Ch.E.J., 18, 31 (1972).

29. R. G. Boothroyd, Transactions of The Institute of Chemical
     Engineers, 44, 306 (1966).

30. S. L. Soo and G. J. Trezek, Ind. Eng. Chem.; Fundamentals, 5,
     388 (1966).

31. W. T. Sproull, Nature, 190, 976 (1961).

32. F. A. Zenz, Ind. Eng. Chem.; 3, 65 (1964).

33. M. Corn, Journal of The Air Pollution Control Association, 11, No. 11 (p. 523) and No. 12 (p. 566), 1966.

34. R. A. Bagnold, The Physics of Blown Sand and Desert Dunes, Methuen, London, 1941.

35. C. Y. Wen and H. P. Simons, A.I.Ch.E.J., 5, 263 (1959).

36. J. H. Holden, U.S. Bureau of Mines, Private Communication, 1967.

37. F. A. Zenz, Ind. Eng. Chem., 41, 2801 (1949).

38. W. J. Carney, Ph.D. Thesis, West Virginia Univ., Morgantown, W. Va., 1954.

39. R. H. Koble, P. R. Jones, and W. A. Kohler, American Ceramic Society Bulletin, 32, 357 (1953).

40. F. A. Zenz and D. F. Othmer, Fluidization and Fluid Particle Systems, Reinhold, New York, 1960.

41. D. Kunii and O. Levanspiel, Fluidization Engineering, Wiley, New York, 1969.

42. D. S. Koons and B. E. Lauer, Ind. Eng. Chem., 53, 970 (1961).

43. C. W. Sandy, T. E. Daubert, and J. H. Jones, "Vertical Dense Phase Gas-Solid Transport," paper presented at the Symposium on Fluidization, Part IV, 64th National A.I.Ch.E. Meeting, New Orleans (March 1969).

44. M. N. Kraus, Chem. Eng., 167-182 (April 12, 1965); and 149-164 (May 10, 1965).

45. M. N. Kraus, Chem. Eng., 59-65 (October 13, 1969).

46. M. N. Kraus, Pneumatic Conveying of Bulk Materials, Ronald Press, New York, 1968.

47. J. D. Spencer, T. J. Joyce, and J. H. Faber, "Pneumatic Transportation of Solids," Proceedings of Institute of Gas Technology - U.S. Bureau of Mines Symposium, Morgantown, W. Va., Oct. 1965, Information Circular No. 3314, U.S. Bureau of Mines, 1966.

48. M. C. Hawk, ed. Bulk Materials Handling, Vol. 1, Univ. of Pittsburgh, Pittsburgh, 1971.

49. BHRA Fluid Engineering, Bedford, England, and Department of Chemical Engineering, City University, London, England, Pneumotransport - I, First International Conference on the Pneumatic Transport of Solids in Pipes, 1971.

50. J. C. Short, Bulk Material Handling, Vol. 1, Univ. of Pittsburgh, Pittsburgh, Pa., 1971.

51. H. G. Gibson, R. L. Peskin, H. A. Dwzen, and J. D. Spencer, ASME Paper No. 66-FE-22, ASME-EIC Fluid Engineering Conference, Denver, Colorado, April 25-28, 1966.

52. C. Y. Wen and A. F. Galli, "Dilute-Phase Systems," in
    Fluidization, J. F. Davidson and D. Harrison, Eds., Academic
    Press, New York, 1971, pp. 677-710.

53. H. G. Gibson and G. E. Fasching, Proceedings of Institute of
    Gas Technology - U. S. Bureau of Mines Symposium, Morgantown,
    W. Va., Oct. 1965, Information Circular No. 8314, U.S. Bureau
    of Mines, 1966, pp. 42-52.

54. K. Cook, Chem. Ind., 1839-1840, December 20, 1968.

55. Anonymous, "Keeps Sugar Free Flowing," Food Eng., 42, 88-89
    (October 1970).

56. E. B. Skubik, Plant Eng., 24, 24 (July 9, 1970).

57. L. J. Schlink, Rock Prod., 73, 52-53 (June 1970).

58. J. L. Konchesky and T. J. George, J. Mining Congr., 57,
    42-46 (December 1971).

59. W. R. Huff and J. H. Holden, Proceedings of Institute of Gas
    Technology - U.S. Bureau of Mines Symposium, Morgantown, W. Va.,
    Oct. 1965, Information Circular No. 8314, U.S. Bureau of Mines,
    1966, pp. 97-111.

60. P. Wellman and S. Katell, Proceedings of Institute of Gas
    Technology - U.S. Bureau of Mines Symposium, Morgantown, W. Va.,
    Oct. 1965, Information Circular No. 8314, U.S. Bureau of Mines,
    1966, pp. 139-145.

61. Anonymous, "Pneumatic Tubes For Garbage Collection Gets U.S.
    Trial Run," Prod. Eng., 41, 16 (March 2, 1970).

62. Anonymous, "Pneumatic Transport of Water Treatment Chemicals",
    Water and Sewage Works, 120, R66-68 (April 30, 1973).

63. Anonymous, "Pneumatic Conveying Practical For Producing
    Granular Gypsum", Modern Materials Handling Engineering, 26,
    147 (November 1971).

64. J. W. Reynolds, Canadian Mining and Metallurgy Bulletin, 65,
    31-36 (July 1972).

Albert Gomezplata and Alan M. Kugelman

Chemical Engineering Department
University of Maryland
College Park, Maryland

I. INTRODUCTION

There are numerous industrial sequences for processing solid-gas systems. Many processing steps involve the use of available equipment designed for other processes. For instance, the drying of powder ores with hot gases in ball mills, or the removal of solid particles from a gas stream by scrubbing with a liquid in a spray or packed tower. Other processing steps might involve equipment specifically designed for gas-solid processing. Examples would be a cyclone separator, an electrostatic precipitator, or a spray coating machine.

Our emphasis will be on processing systems involving moving beds, spouting beds, or fluidized beds. Major attention will be given to the fluidized bed because it is the most flexible and the most widely used of the three mentioned processing systems. Fluidized beds have received considerable attention during the last decade, and their design can be undertaken with a high degree of reliability.

A fluidized bed is simply a bed of solid particles with a stream of gas passing upward through the particles at a high enough rate to set them in motion. Figure 1 shows three fluidized beds exhibiting a range of fluidizing conditions.

The gas velocity in case (a) is just above that required to expand the bed of solid particles (minimum fluidization velocity). At a greater gas velocity, gas bubbles appear and penetrate through the

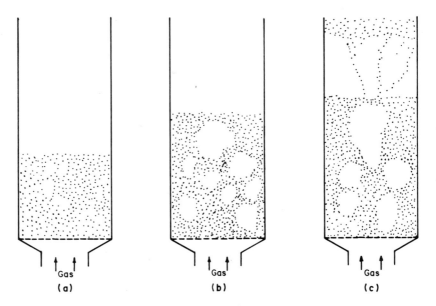

Fig. 1. Range in fluidizing conditions. (a) Smooth fluidization; (b) bubbling fluidization; (c) slugging fluidization.

expanded bed as shown in case (b). The solids in this type of bed
undergo violent agitation due to the rising gas bubbles. Greater
gas velocities give rise to some gas bubbles that occupy the entire
cross section of the column and are known as gas slugs. The breakup
of slugs at the top of the bed causes considerable solid spouting
and some carry-over of solid in the exit gas stream. Still greater
velocities lead to the pneumatic transport of the solids in the gas
stream. Some of the most desirable characteristics of a fluidized
bed are:

(1). The vigorous solid movement, good mixing, and solid temperature; properties tend to be uniform in the bed.

(2). The rate of heat transfer between the fluidized bed and
heating surfaces is increased substantially over what it would be
for the gas alone and the transfer surfaces.

(3). The large surface area of the solid particles makes the
fluidized bed desirable for mass transfer and chemical reactions
involving the gas-solid interfaces.

(4). The ability to withdraw streams, with good stream flowability, is a major advantage. In this respect the fluidized bed
resembles a fluid bed. An example of a fluidized bed with inlet and
outlet streams is shown in Fig. 2, for a fluid catalytic cracking
system.

Some limitation of a fluidized bed are:

(1). The solid particles must have proper fluidizing characteristics and may undergo undesirable attrition as well as elutriation
(carry over).

(2). The gas throughout is limited by the relatively low velocities needed for pneumatic transport unless the equipment design
makes provision for the pneumatic transport.

(3). The uniform conditions in a fluidized bed are usually undesirable for mass transfer operations.

The spouted bed is a combination of a dilute fluidized phase and
a coexistent moving bed of solids. In this respect it offers the

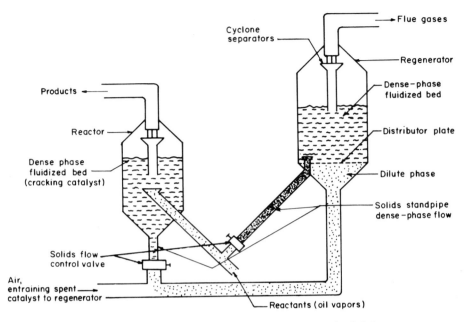

Fig. 2. Fluidized bed showing ability to inject and withdraw streams.

attributes of slowly moving beds while maintaining many desirable aspects of fluidized systems. Figure 3 is a schematic diagram of a spouted bed [1]. Gas enters the apparatus through an opening at the bottom of the conical section at a sufficient velocity to form a central channel in which solids are carried upward. The solids-gas ratio in the central channel is of the same order of magnitude as that found in a typical dilute-phase fluidized system [1]. At the top of the channel, solids spill over radially and form an annular, downward moving stream of solids. The solids concentration in the channel increases with bed height; the gas distribution over the column cross section varies with bed height. Thus, a spouted bed consists of a central gas spout carrying particles upward, and a downward moving stream of particles through which some gas is flowing countercurrently [2].

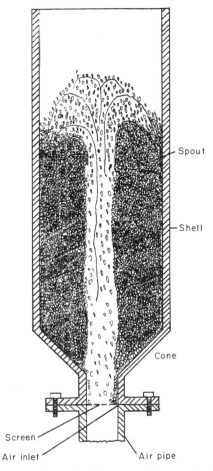

Fig. 3. Schematic diagram of a spouted bed.

A moving bed is the operation in which solid particles, which are
in a packed bed, are made to contact a gas stream. The particles can
be conveyed through the gas stream or slowly move down the bed by
gravity. An example would be the upper part of a blast furnace. The
use of a supported moving bed is a recourse to be used in the case
of a nonfluidizable or poorly fluidizable solids processing system.

The choice of using a fluidized bed, spouting bed, or a moving bed generally depends on the flowability and other characteristics of the solids to be contacted. (Refer to Chapter 2 by R. L. Carr for the required and desirable characteristics of the solids.) In some cases it is possible to modify the solids to give them the required characteristics for the most efficient operation. An example would be the ability to grind solids to a desirable particle-size distribution for fluidization. However, in other cases the choice of operation is dictated by the characteristics of the solid. If a very fine powder that has poor fluidization characteristics is to be processed, there might not be any alternative but to use a moving bed or some other specialized contactor. Chapter 1 deals with the characteristics of solids and should be studied to characterize the flowability and other solids properties. Of course, small-scale laboratory tests are an expedient way to erase doubts about the characteristics of the solids involved. This chapter proceeds with the understanding that solid characteristics have been checked.

One of the most desirable properties of a fluidized operation is easy handling of solid-gas streams to or from the process. In general, if a solid has desirable fluidization characteristics it can be handled as a stream in a fluidized state or by pneumatic transport.

II. FLUIDIZED BED

In discussing the design characteristics, we concentrate on the key design parameters needed for consideration in a fluidized-bed application. These include energy requirements, physical parameters such as the diameter and height of the bed, and expected elutriation and quality of fluidization. Basic fluidized-bed parameters such as bed void fraction and minimum fluidization velocity are needed to predict the foregoing quantities and to design the support plate, feeder, and possibly the collector system.

A. BASIC PARAMETERS

In order to obtain a fluidized bed, a fluid, either a liquid or a gas, is passed upward through a bed of small solid particles supported in a vertical container by a grid. At certain fluid velocities, the fluid will support the particles, giving them freedom of mobility, and yet not entraining them overhead. Such a fluidized bed resembles a vigorously boiling fluid with the solid particles undergoing extremely turbulent motion. The amount of turbulence increases with fluid velocity.

Superficial velocities, based on the total cross-sectional area, are usually utilized in dealing with fluidized-bed operations. A superficial velocity of 5 ft/sec is obtained when passing 0.5 cu ft/sec of a fluid through a bed with a total cross-sectional area of 0.1 sq ft.

A typical pressure drop versus superficial velocity curve is shown in Fig. 4. At low gas velocities the bed of particles is essentially a packed bed, and the pressure drop is proportional to the superficial velocity. As the gas velocity is increased, a point is reached at which the bed behavior changes from fixed particles to suspended particles. At this point a bed expansion occurs. The superficial velocity required to first suspend the bed particles is known as the minimum fluidization velocity and is denoted by the symbol umf. The slope of the pressure drop curve decreases after the minimum fluidization velocity is reached, with only a gradual increase in pressure drop with velocity up to the point where entrainment of the particles takes place. The pressure-drop curve in the vicinity of the minimum fluidization velocity is not unique, but depends on the packing (position) of the packed bed of particles. At the incipient point of fluidization the pressure drop of the bed will be very close to the weight of the particles divided by the cross-sectional area of the bed (W/A).

The increase in pressure drop above this datum depends on the properties of both the solids and the gas, but in general does not exceed 20% of the pressure drop caused by the weight of the bed.

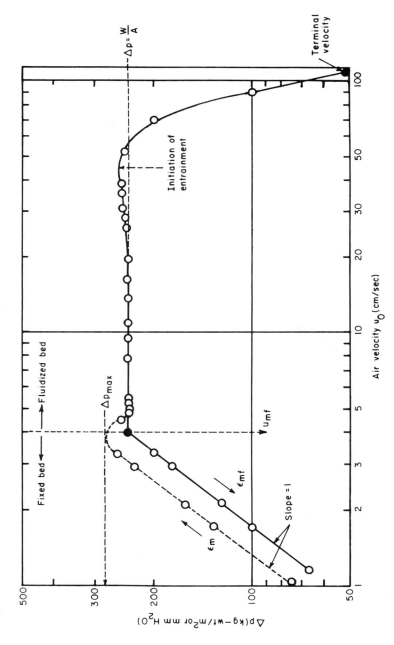

Fig. 4. Typical pressure drop curve as a function of gas velocity.

The minimum fluidization velocity and the approximate pressure drop of a fluidized (W/A) are both very useful parameters. The minimum fluidization velocity sets the lower limit of possible operating velocities and the approximate pressure drop can be used to approximate pumping energy requirements.

The initial suspension of particles at incipient fluidization can be described by a balance of forces which describes the minimum fluidization conditions:

$$\Delta P_{mf} \ A = (1 - \varepsilon_{mf})(\rho_p - \rho_g)U_o \frac{g}{g_c}$$

where ΔP = pressure drop, A = cross-sectional area, ε_{mf} = minimum fluidization void fraction, ρ_p = density of particles, ρ_g = density of gas, U_o = superficial gas velocity, and g/g_c = ratio of gravitational acceleration and gravitational conversion factor.

For the normal gas fluidized bed the density of the gas is much less than the density of the solids and the balance of forces can be reduced to

$$\Delta P_{mf} = \frac{W}{A}$$

where

$$W = (1 - \varepsilon_{mf})\rho_p \frac{g}{g_c}$$

is the weight of the particles in the bed. We previously mentioned that the pressure drop of a fluidized bed is approximately the weight of particles per unit cross-sectional are of bed.

The minimum fluidization void fraction (ε_{mf}) can best be estimated from experimental data of loosely packed beds. Table 1 shows typical data as a function cf particle diameter and sphericity (ϕ_s).

TABLE 1

Typical Void Fraction of Packed Beds as a Function of
Particle-to-Container Diameter Ratios[a]

Particle shape	Void fraction at particle-to container diameter ratios of				
	0.10	0.20	0.30	0.40	0.50
Uniform spheres	0.37	0.41	0.46	0.50	0.54
Mixed spheres	0.33	0.36	0.39	0.42	0.45
Uniform cylinders	0.35	0.40	0.39	0.52	0.56
Granules	0.48	0.53	0.60	-	-

[a]Source: M. Leva, "Fluid Flow Through Packed Beds,"
Chem. Eng., 56, 116 (1949).

The sphericity of the particle is used to characterize nonspherical
particles and is defined as

$$\phi_s = \frac{\text{surface of sphere}}{\text{surface of particle}} \text{ (of same volume).}$$

The sphericity of spheres is equal to 1.

To estimate the minimum fluidization velocity the relationships
suggested by Wen and Yu [3] are recommended.

For small particles where Re < 20

$$U_{mf} = \frac{d_p^2(\rho_s - \rho_g)g}{1,650\mu}$$

and for large particles where Re > 1,000

$$U_{mf} \sqrt{\frac{d_p(\rho_s - \rho_g)}{24.5\rho_g}}$$

For other Reynolds numbers the following expression has to be solved
to estimate the minimum fluidization velocity:

$$U_{mf} = \frac{\mu}{d_p \rho_g} \left[\left[33.7^2 + \frac{0.0408 d^3 \rho_g (\rho_s - \rho_g) g}{\mu^2} \right]^{1/2} - 33.7 \right]$$

If the sphericity and minimum void volume are known more rigorous
expressions can be used [4].

B. OPERATING CONDITIONS

The actual operation of a fluidized bed at predetermined operating conditions is still difficult to predict [4]. Much progress has
been made recently in the understanding of the behavior of single
gas bubbles rising through a fluidized bed [4], but the complex interaction of a multiple of bubbles and the fluidizing medium cannot be
accurately predicted. Most of the experimental studies to date have
been on 2- to 6-in. diameter beds, and it is for this range in diameters that sufficient experimental data are available to characterize
operating conditions. For larger diameters, judicious care is needed for scale-up; It is recommended that the literature be consulted
for engineering data for large-scale processes. The most critical
part of a fluidized-bed design is the bed support or gas distributor
system. The design of the gas distributor will be discussed in part
3 of this section.

1. Superficial Velocity

The minimum fluidization velocity sets the lower limit of operating velocities. The upper limit is set by the pneumatic transport
velocity (see Chapter 3). Actually the superficial velocity has to
be two to five times higher than the minimum before the desirable
characteristics of a well-mixed, turbulent bed are obtained. A more
practical upper limit is usually established by the amount of carry-
over of fines that the process or equipment can handle. A cyclone

separator that can recycle carry-over allows higher operating velocities for a fluidized bed. To evaluate the limiting velocity of
carry-over solids, the terminal velocity of the smallest fines should
be used. Methods to estimate the terminal velocity of single particles can be found in many texts [1,2,4]. Figure 5 from Pinchbeck
and Popper [5] shows the experimental ratio of terminal velocity to
minimum fluidization velocity for different-sized particles. A minimum ratio of 9:1 and a maximum of 90:1 is possible depending on particle size. Large-diameter beds can operate at even higher ratios
if part of the gas is allowed to short circuit or to channel through
the bed. In general, though, a poor distribution of gas throughout
the bed is not desirable.

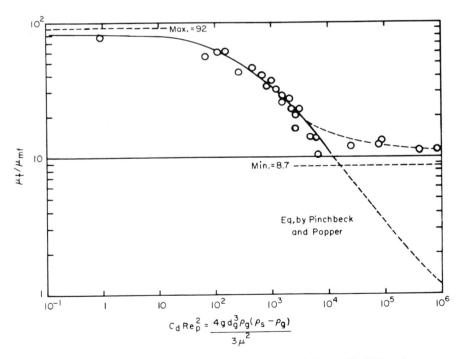

Fig. 5. Ratios of terminal velocity to minimum fluidization
velocity.

2. Quality of Fluidization

Gas beds are inherently harder to fluidize than liquid beds because of the larger density difference between the fluidizing medium and the solids. With a liquid a uniform suspension of particles, which is termed particulate or dispersed fluidization, is possible. With a gas usually colonies or aggregates of particles circulate in the bed, and gas pockets or bubbles rise through the bed. This is called aggregate fluidization and is characteristically not as uniform as particulate fluidization.

Four dimensionless groups have been identified to characterize the uniformity of fluidization [6,7]:

$$F_{r_{mf}} = \frac{U_{mf}^2}{d_p g}$$

$$Re_{mf} = \frac{d_p U_{mf} \rho_g}{\mu}$$

$$\frac{\rho_s - \rho_g}{\rho_s} \quad \text{and} \quad \frac{L_{mf}}{dt}$$

An increase in the value of these groups signifies particulate and aggregate fluidization:

$$(F_{r_{mf}})(Re_{mf})\left(\frac{\rho_s - \rho_g}{\rho_g}\right)\frac{L_{mf}}{dt} < 100 \quad \text{particulate}$$

and

$$(F_{r_{mf}})(Re_{mf})\left(\frac{\rho_s - \rho_g}{g}\right)\frac{L_{mf}}{dt} > 100 \quad \text{aggregate}$$

In general gas-solid fluidization is aggregate and exhibits gas bubbles rising through the bed. The closer the value of the product of the terms above to particulate fluidization the better the expected uniformity of the fluidized bed.

The relative uniformity and behavior of an aggregate fluidized bed ranges from a uniform suspension with evenly distributed small bubbles, to a vigorously boiling bed with large bubbles passing through and eruptions at the surface, to an extreme of having slugs of gas that are the same diameter as the column rising through the column and imparting a periodic upheaval of the entire contents of the bed. For large-diameter beds this slug regime will be replaced by local points of large bubble activity, or channeling, and other areas of the bed with no bubble activity. The three most critical factors that control this quality of fluidization are the particle size, particle-size distribution, and the gas distribution used.

It is usually desirable to have a distribution of particle diameters to obtain a uniform fluidized bed [1,2]. Beds with a particle size diameter distribution between 20 and 80μ generally give the best fluidization results. Larger-diameter particles can be tolerated as long as there is a suitable distribution of particle sizes. Very small particles tend to be carried over by the gas. In the case of beds composed of small particles less than 80μ it is not possible to fluidize the mixture without gas channeling and slugging [8]. The effect of operating parameters on elutriation is reviewed by Leva [1].

The importance of the gas distributor on the quality of fluidization has been reported by Grohse [9], and Fig. 6 shows the marked effect that a distributor can have on bed expansion. A screen distributor has the lowest pressure drop and the worse fluidization characteristics. A porous plate has the highest pressure drop but gives the best quality of fluidization. An orifice plate is between the two from either a pressure drop or a quality of fluidization standpoint and is the most likely to be used in a large fluidized bed.

Many schemes have been proposed to measure the quality of fluidization [2]. Three practical methods involve the measurement of pressure drop, capacitance, and light transmission fluctuations. Smooth aggregate fluidization is characterized by small amplitude, high frequency fluctuations, while slugging shows up as large amplitude, low frequency oscillations [10]. Figure 7 shows typical pressure drop versus velocity relations for channeling and slugging fluidized bed.

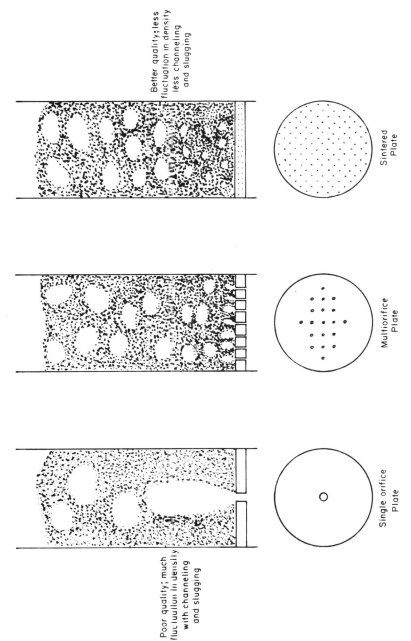

Fig. 6. Effect of gas distributor on quality of fluidization.

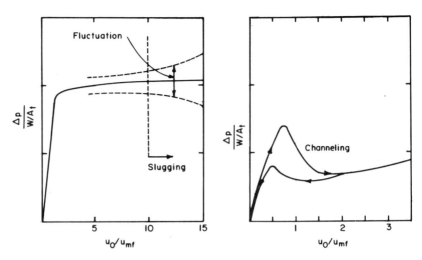

Fig. 7. Pressure drop for channeling and slugging fluidized beds.

3. Gas Distributor

In laboratory studies the most common gas distributor has been a
screen, multiorifice plate, or a porous metallic or ceramic plate.
As mentioned before, the best fluidizing conditions are obtained with
a porous plate, but at the expense of a high pressure drop. A screen
distributor, on the other hand, has a low pressure drop but the quality of fluidization is usually poor. An orifice plate distributor can
be designed to approach the gas distribution characteristics of a
porous plate or a screen. In general, the higher the pressure drop
of the orifice plate, the better its fluidization characteristics.

In industrial applications that require large-diameter beds, a
porous-plate distributor is not practical because of fabrication
problems, high pressure drop, and plugging by fine particles. Distributor designs involving multiorifice plates or an array of grates
or bars are then necessary. Figure 8 shows some typical industrial
distributor designs. The multiorifice design can be either flat or

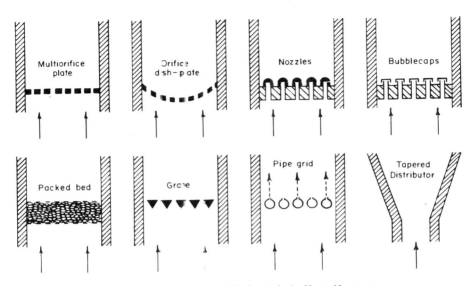

Fig. 8.   Examples of industrial distributors.

concave for greater strength.   Multinozzle or bubble-cap designs have
been used and have the advantage of preventing the solids from falling
through the distributor; however they are more expensive to fabricate.
A packed section has also been used successfully because it can be
designed to have a reasonable pressure drop and the packing tends to
prevent the leakage of solids particles through the distributor.   Low-
pressure-drop distributors consist of metal grates, pipe grids for
simultaneous heating or cooling, and in some cases a tapered distributor with no support plate or grid.   Low-pressure drop distributors
can result in a poorly fluidized bed but provide for a maximum gas
through-put at a minimum pressure drop.   In these cases the use of
baffles to improve the quality of fluidization has been reported [4].

Kunii and Levenspiel [4] review the suggested design criteria for
a good gas distributor and list the three conditions that determine
the minimum recommended pressure drop:

1.   The pressure drop across the distributor must be at least 10% of the pressure drop across the bed.

2.   The pressure drop across the distributor must be at least 100 times the expansion loss when flow passes from the inlet connection into the vessel.

3.   The pressure drop across the distributor must be at least equal to 35 cm of water.

If the pressure drop indicated by the foregoing criteria can be tolerated, a high pressure-drop distributor is recommended.   If a low pressure-drop distributor is necessary the use of baffles to help the quality of fluidization should be considered.   Table 2, from Levenspiel [4], summarizes successful industrial distributors and their pressure-drop characteristics.

An orifice-plate distributor can be designed by using the following procedures [4].

1.   Determine the necessary pressure drop across the distributor from the design criteria reviewed earlier.

2.   Use Fig. 9 to evaluate the orifice coefficient C_d' from the calculated Reynolds number.   An open area in the distributor of 10% is assumed.

3.   Evaluate the velocity of fluid through the orifices at the approach density and temperature.

$$U_{or} = C_d' \left(\frac{2g_c \ \Delta P}{\rho_g} \right)$$

The ratio U_o/U_{or} is the fraction of open area in the distributor plate.

4.   Balance the number of orifices N_{or} and the corresponding orifice diameter using the expression,

$$U_o = \frac{\pi}{4} d_{or} U_{or} N_{or}$$

TABLE 2

Design Data for Pressure Drop Across Distributors

Case and type of distributor	Designed ΔP_d (g-wt/cm^2 or cm H$_2$O)
Drying of fine coal within a shallow fluidized bed, multiorifice plate U_o = 3.7-4.3 m/sec	Open area 4.5% ΔP_d = 46 cm H$_2$O
FCC[a] reactor and regenerator, orifice dish plate d_t = 4-12m	Orifice diameter 3.8-5.1 cm ΔP_d = 35-70 cm H$_2$O
Roasting of sulfide ore, nozzles L_f = 1-1.5m nozzle type	Nozzle diameter 0.4-0.6 cm ΔP_d = 25-30 cm H$_2$O for roasting of ordinary sulfide ore ΔP_d = 30-50 cm H$_2$O for roasting of nickel sulfide ore
Experimental air fluidized beds of large diameters multi-orifice plate	

Open area (%)	7.3	0.4	0.5
$d_p(\mu)$	65	65	-
d_t(m)	0.601	2.14	3.96
L_f(m)	1.68	6.1	8.38
U_o(cm/sec)	14.6-28.7	10.7	31.4

[a]Fluid catalytic cracking.

Source: Kunii and Levenspiel, Fluidization Engineering, 87, Wiley, 1969.

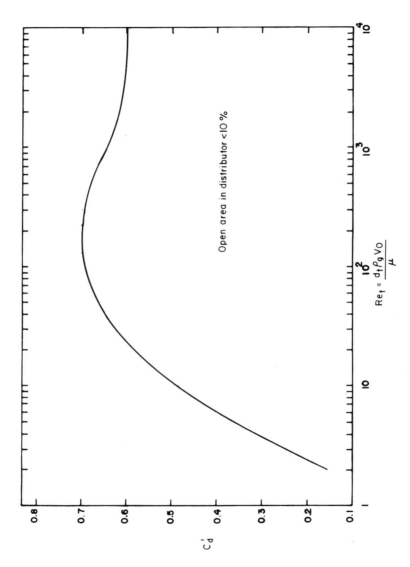

Fig. 9. Orifice coefficient versus Reynolds number based on diameter of approach chamber [4].

A choice of number of orifices and corresponding orifice diameter are
then made available to meet the pressure-drop design criteria.

4. Void Fraction, Bed Height, and Bed Diameter

There are as yet no reliable methods to predict the void fraction
or bed height of a fluidized bed at operating conditions. The bed
height and bed void fraction are related by the following expression,

$$\frac{L_f}{L_{mf}} = \frac{1 - \varepsilon_f}{1 - \varepsilon_{mf}}$$

Figure 10 shows the range of bed height ratios that can be obtained
as a function of gas velocity for the case of 100μ particles as

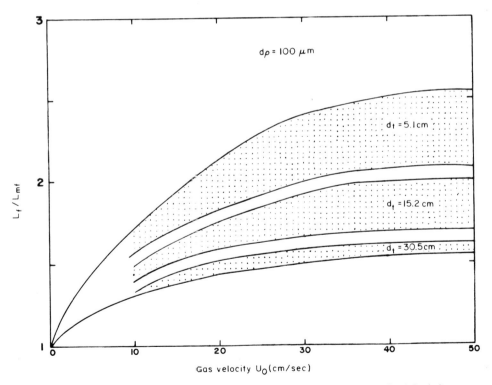

Fig. 10. Effect of bed diameter and gas velocity on bed height
expansion ratio [4,11].

reported by Zenz and Othmer (11). The large range of bed expansion
found for each bed diameter is due to oscillations in bed height.

The bed diameter is usually determined by the operating superficial velocity and the volume rate of gas that must be handled. The
bed cross-sectional area is,

$$A = \frac{Q}{U_o}$$

The bed diameter can be calculated from the bed cross-sectional area
since

$$A = \frac{\pi D_c^{\ 2}}{4}$$

The operating superficial velocity is usually the maximum that can be
handled without excessive elutriation.

5. Energy Requirements

The energy requirements for a given fluidized-bed operation are
equal to the sum of the energy requirements of the physical and chemical steps in the process plus the energy requirements for pumping
the gas through the bed. An approximate estimation of pumping requirements can be obtained from the following sequence of calculations.

1. Approximate pressure drop through bed: $\Delta P = \frac{W}{A}$.
2. Approximate pressure drop through distributor: $\Delta P_d = 0.1\ \Delta P$.
3. Approximate total pressure drop: $\Delta P_T = \Delta P + \Delta P_d$.
4. Approximate pumping requirements: $E = U_o A\ \Delta P_T$.

A more accurate estimate can be obtained by using calculation methods
found in the texts by Kunii and Levenspiel (4) and Orr (2).

C. HEAT, MASS TRANSFER, AND REACTION

Perhaps the most useful property of a fluidized bed is its re-
markable temperature uniformity. The fast recirculation of solids,
due to gas bubbling, results in effective thermal conductivities up

to one hundred times that of silver in both radial and axial directions. Heat transfer rates to containing walls or to internal surfaces are much higher in a fluidized bed than they are for gas flow alone through the bed and about ten times greater than for a packed bed. Heat transfer coefficients in the range of 5 to 900 Btu/hr ft^2 °F have been reported with values in the range of 50 to 100 Btu/hr ft^2 °F being common [2].

The heat transfer coefficient is found to increase with gas velocity after the onset of fluidization and to pass through a maximum as the gas velocity is increased. This maximum is reached at a point where the increase in gas velocity does not appreciably increase the solid turbulence in the bed, but where it does result in a greater bed expansion.

Poor fluidization will decrease the heat transfer coefficient as bed height is increased . The use of baffles has been reported [2] as a method to improve the fluidization and to counter this decrease in the rate of heat transfer as the bed height is increased.

Numerous correlations can be found in the literature to predict heat transfer coefficients in fluidized beds. A summary of correlations can be found in standard texts [1,2,4]. The predictions from these correlations can differ narkedly. It is recommended that a range of coefficients be estimated, when appropriate, and more than one correlation be used.

The heat transfer coefficient (h) is defined by the rate equation,

$$q = hA \ \Delta T$$

where q = rate of heat transfer (Btu/hr), A = heat transfer surface area (ft^2), and ΔT = mean temperature difference between fluidizing medium and heat transfer surface.

1.   Heat Transfer Between the Wall and the Fluidized Bed

Some of the correlations available for predicting, the heat transfer coefficient between the container wall and the fluidized bed are listed below.

1. Toomey and Johnstone [12]:

$$\frac{hd_p}{k_g} = 3.75 \left(\frac{d_p \rho_g U_{mf}}{\mu} \log \frac{U_o}{U_{mf}}\right)^{0.47}$$

2. Levenspiel and Walton [13]:

$$\frac{hd_p}{k_g} = 0.6 \left(\frac{C_{pg}\mu}{k_g}\right)\left(\frac{d p \rho_g U_o}{\mu}\right)^{0.3}$$

3. Dow and Jakob [14]:

$$\frac{hd}{k_g} = 0.55 \left(\frac{d_p \rho_g U_o}{\mu}\right)^{0.8} \left(\frac{\rho_s C_{ps}}{\rho_g C_{pg}}\right)^{0.25} \left(\frac{d_t}{d_p}\right)^{0.03} \left(\frac{L_f}{d_t}\right)^{-0.65} \left(\frac{1 - \varepsilon_f}{\varepsilon_f}\right)^{0.25}$$

4. van Heerden et al. [15]:

$$\frac{hd_p}{k_g} = 0.58 \left(\frac{C_{pg}\mu}{k_g}\right)^{0.5} \left(B \frac{d_p \rho_g U}{\mu}\right)^{0.45} \left(\frac{C_{ps}}{C_{pg}}\right)^{0.36} \left[\frac{\rho_s(1 - \varepsilon_{mf})}{\rho_g}\right]^{0.18}$$

where the factor B has the following values: carborundum 0.62 to 0.78, coke 0.39 to 0.58, and ferric oxide 0.59.

5. Wen and Leva [16] as rearranged in reference 2:

$$\frac{hd_p}{k_g} = 0.16 \left(\frac{C_{ps} \rho_s d_p^{1.5} g^{0.5}}{k_g}\right)^{0.4} \left(\frac{d_p U_o g}{\mu}\right)^{0.36} \left(\frac{E}{R}\right)^{0.36}$$

where E and R are fluidization parameters and can be evaluated from Table 3 as a function of particle diameter and the ratio of G_g, the

TABLE 3

E and R as Functions of the Fluidization Mass Velocity

Ratio G_g/G_{mf} [a]

G_g/G_{mf}	Particle Diameter (in.)											
	0.002		0.004		0.006		0.008		0.010		0.012	
	E	R	E	R	E	R	E	R	E	R	E	R
1	0	1	0	1	0	1	0	1	0	1	0	1
2	0.40	1.06	0.32	1.07	0.23	1.10	0.18	1.12	0.15	1.16	0.11	1.21
4	0.66	1.12	0.55	1.15	0.45	1.19	0.33	1.28	0.24	1.39	–	–
6	0.76	1.16	0.66	1.20	0.56	1.27	0.42	1.41	–	–	–	–
8	0.81	1.19	0.71	1.24	0.62	1.33	–	–	–	–	–	–
10	0.84	1.21	0.75	1.28	0.65	1.39	–	–	–	–	–	–
12	0.87	1.24	0.77	1.33	–	–	–	–	–	–	–	–
14	0.89	1.26	0.80	1.36	–	–	–	–	–	–	–	–
16	0.90	1.28	–	–	–	–	–	–	–	–	–	–

[a] From C. Y. Wen and M. Leva, "Fluidization-Bed Heat Transfer: A Generalized Dense-Phase Correlation," A. I. Ch. E. J., 2, 488 (1956).

gas mass velocity of fluidization, to G_{mf}, the gas mass velocity at incipient fluidization.

The correlation of Wen and Leva is based on data from a number of studies and covers a wide range of solids. The equation predicts 95% of the data used to within $\pm$ 50%.

2. Heat Transfer Between Internal Tubes and the Fluidized Bed

a. **Vertical Tubes.** Wender and Cooper [17] correlate the data of a number of investigators for vertical tubes with the expression,

$$\frac{hd_p}{k_g} = 0.01844 \ C_R(1 - \epsilon_f)\left(\frac{C_{pg}\rho_g}{k_g}\right)^{0.43} \left(\frac{d_p\rho_g U_o}{\mu}\right)^{0.23} \left(\frac{C_{ps}}{C_{pg}}\right)^{0.8} \left(\frac{\rho_s}{\rho_g}\right)^{0.66}$$

for a range of $\dfrac{(d_p\rho_g U_o)}{\mu}$ of from 0.01 to 100. The group $(C_{pg}\rho_g/k_g)$ has units of sec/cm^2 and the factor C_R can be found from Fig. 11. A deviation of $\pm 20\%$ is reported for the 323 data points used.

b. **Horizontal Tubes.** Vreedenberg [18,20] recommends the following correlations for horizontal tubes in fluidized beds:

$$\frac{hd_t}{k_g} = 0.66 \left(\frac{C_{pg}\mu}{k_g}\right)^{0.3} \left[\left(\frac{d_t\rho_g U_o}{\mu}\right)\left(\frac{\rho_s}{\rho_g}\right)\left(\frac{1 - \epsilon_f}{\epsilon_f}\right)\right]^{0.44} \quad \text{when} \quad \left(\frac{d_t\rho_g U_o}{\mu}\right) < 2,000$$

and

$$\frac{hd_t}{k_g} = 420 \left(\frac{C_{pg}\mu}{k_g}\right)^{0.3} \left[\left(\frac{d_t\rho_g U_o}{\mu}\right)\left(\frac{\rho_s}{\rho_g}\right)\left(\frac{\mu^2}{d_p^3\rho_s P}\right)\right]^{0.3} \quad \text{when} \quad \left(\frac{d_t\rho_g U_o}{\mu}\right) > 2,500$$

The correlations include data for large-scale beds.

3. Mass Transfer and Chemical Reaction

There are two major types of industrial operations that involve mass transfer in fluidized beds. One is the drying of wet solids, which involves the introduction of a slurry into the bed and then a

Center of vessel   Wall of vessel

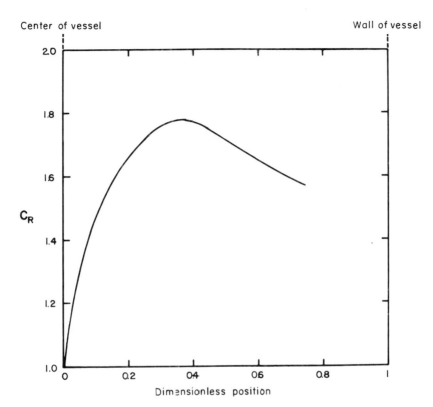

Fig. 11.   Correction factor C_r for nonaxial location of immersed tubes.

simultaneous transfer of heat and mass from the individual particles. The Atomic Energy Commission has found this method to be effective for treating radioactive solid wastes [21].   Details on fluidized bed dryers can be found in review articles [22] and texts [23,24].

When a chemical reaction takes place in a fluidized bed, the operation will also involve mass transfer.   Examples are the catalytic cracking of petroleum, and SO_2 removal from an air stream via a limestone process.   The operation involves interaction between the incoming gas, which is usually a reactant, and the solid particles, which act as reactants or as catalysts.   The dominant regions where mass transfer and reaction take place are in the gas bubbles and the immediate vicinity of the bubbles. The understanding of the bubble phenomenon has led to the successful modeling of the simultaneous process of mass transfer and chemical reaction.

Details of the basic bubble model, proposed by Davidson, that account for the movement of both gas and solids can be found in the texts by Davidson and Harrison [25]. Excellent summaries are also avialable [4,24,27]. Figure 12 shows the rise of a single NO_2 bubble through a two-dimensional bed. The most interesting and useful observations are that the gas in the bubble stays in the vicinity of the bubble, that solids can circulate through the gas pocket in the bubble, and that there is a cloud of solid that rises with the bubble.

Fig. 12.   Rise of a NO_2 bubble through a two-dimensional bed [25].

The principal properties of the gas bubble that have been determined are the bubble gas velocity as a function of bubble diameter
and the mass transfer coefficients between the gas bubble, the cloud
surrounding the bubble, and the dense phase of the fluidized bed.
Using the foregoing characteristic of the gas bubbling process, Kunii
and Levenspiel [4] proposed a bubbling bed model in order to predict
the extent of a chemical reaction in a fluidized bed. The only adjustable parameter in the model is the effective size of the bubbles.
Kato and Wen [27] have more recently proposed the "bubble assemblage
model," which is geared for computer simulation and has adjustable
parameters. In this model the fluidized bed is considered to be a
series of compartments, the height of which depends on the bubble size
predicted for that height. Experimental conversions predicted by the
model are shown in Fig. 13 together with a table of the available experimental data used to test the model. The "bubble assemblage model"
is the most comprehensive one presently available in the literature.
Unfortunately all the experimental data available to test the model
were obtained in small-diameter (2 to 8-in.) reactors, and the applicability of the model in predicting the performance of large-size
equipment has not been demonstrated. A few references are available
on the problems of the scale-up of fluidized beds [24,28,29] but a
pilot-plant study is still needed for a large-scale plant design.

D.   INDUSTRIAL APPLICATIONS OF FLUIDIZATION

Industrial applications of fluidization are widespread and extremely important. Fluidization principles have been applied to
equipment designs to accomplish both physical and chemical changes.
For example, designs for such physical operations as drying, mixing,
particle sizing, heat exchange, freezing, and adsorption have been
developed that include fluidized, solid phases. Likewise, fluidized
solid catalyst phases are utilized in such chemical processes as the
production of petroleum products and the manufacture of important
organic chemicals. In addition, fluidized beds are also utilized

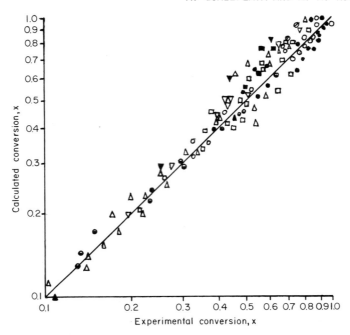

Symbol	Author	d_p(cm)	ρ_p(g/cm^3)	D_R(cm)	L_{mf}(cm)	k(1/sec)
O	Lewis *et al.*	0·0122	0·93	5·2	24 ~ 46	1·4 ~ 8·7
●	Kobayashi *et al.*	0·0194	1·25	8·3	19 ~ 100	0·8 ~ 0·2
△	Massimilla and Johnstone	0·0105	2·06	11·4	18·1 ~ 54	0·0707
◒	Shen and Johnstone	0·00806	1·91	11·4	27 ~ 54	0·022 ~ 0·0068
□	Kobayashi *et al*	0·0194	1·25	20·0	34 ~ 67	0·2 ~ 1·4
▼	Orcutt *et al.*	0·0038 ~ 0·0042	1·14	15·25 ~ 10·15	61 ~ 29·4	0·2 ~ 3·0
▲	Mathis and Watson	0·0103	1·2	8·07	10 ~ 50	0·65
Ø	Echigoya *et al.*	0·0122	1·15	5·3	5 ~ 17	4·0 ~ 8
■	Gomezplata and Shuster	0·0105	0·98	7·6	4 ~ 20	0·775
▽	Ishii and Osberg	0·0088	1·65	15·2	4·5 ~ 26	0·31 ~ 1·5

Fig. 13. Comparison of experimental with calculated conversion for Bubble Assemblage Model [26].

in the metallurgical industries, in the production and upgrading of fertilizers, and in the manufacture of cement.

Quite recently, fluidized-bed processes have been proposed for the conversion of municipal refuse and waste to fuel gas, and for

industrial waste liquor treatment. In addition, fluidized coal gasification processes have been under study and appear to be extremely
promising and economical for converting coal to gaseous fuels. It
thus appears certain that fluidized-bed technologies will continue
to evolve and improve and remain quite important in future years.

The following discussion is intended for the illustration of the
diverse and important applications of fluidization technology in the
chemical process industries today. It is by no means a complete or
exhaustive discussion.

1. Petroleum and Petrochemical Industries

Perhaps the most important application of fluidization technology
found in industry today is the fluid catalytic cracking (FCC) process, which converts heavy gas oils to gasoline and fuel oil. The
primary problems associated with the cracking process are: (1)
supplying the necessary heat to sustain the endothermic cracking
reaction, and (2) the continuous reactivation of the catalyst, which
deactivates extremely fast. The FCC process was developed to overcome these problems and today its primary purpose is to meet the
ever increasing worldwide demand for gasoline.

A typical FCC is shown schematically in Fig. 14. In order to
continuously reactivate the deactivated catalyst, the catalyst is
continuously recirculated in a fluidized state from the reactor,
where the reaction takes place and the catalyst is deactivated by
coke deposits, to the regenerator where the coke deposits are burned
off with air. Since the coke-burning reaction is very exothermic,
the sensible heat necessary to maintain the endothermic cracking
reaction is supplied by the recirculated, fresh catalyst.

In the early 1960's, new zeolite cracking catalysts (the so-
called molecular sieve catalysts) were developed, having activities
much higher than the conventional silica-alumina cracking catalysts.
As the new catalysts replaced the conventional catalysts in the old
units, it became apparent that most of the cracking was taking place
in the short time that the hot, freshly activated catalyst and the
hot feed traveled through the riser [30]. This realization led to

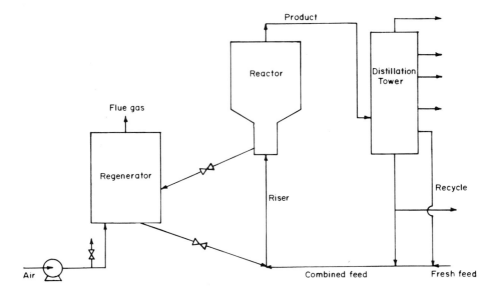

Fig. 14. A typical fluid catalytic cracking plant.

a major recent development in FCC design, the so-called transfer
line or riser cracking. A typical riser plant is shown in Fig. 15
[30]. The reactor is replaced by a catalyst disengager; the desired
reaction occurs in the riser.

Attempts have been made to apply the same fluidization techniques
to other petroleum processes. One such application is fluid catalytic reforming [31]. Catalytic reforming converts low-octane gasoline boiling-range naphthas to high octane gasoline. The major
reactions are extremely endothermic and are accompanied by coke deposition. This suggests fluidization of the catalyst and continuous
regeneration in a regenerator; the heat produced by regeneration is
transferred to the reactor (via the recirculated catalyst) to sustain the endothermic reforming reactions. Because the catalyst de-
cay occurs much less rapidly in the reforming process, fixed-bed
processes [32] and semiregenerative processes [33] are preferred to
the fluidized process.

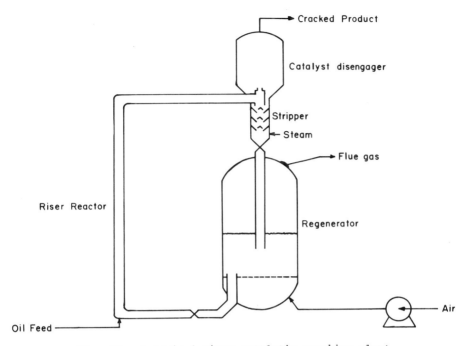

Fig. 15.   A typical riser catalytic cracking plant.

Coking is a high-pressure, high-temperature process that completely converts residual stocks to lighter (gas, some gasoline, and
gas oil) and heavier (coke] materials. A simplified diagram of the
fluidized coking process is shown in Fig. 16 [34]. Steam is used
to fluidize the coke bed in the reactor. The coke is built up on
fluidized particles until a size suitable for removal is obtained;
finer particles are recycled to the reactor.

The application of fluidization to the production of petro-chem-
icals is also very important. This is exemplified by a BASF fluid-
dized-bed process for producing olefins from crude oil [35]. It is
reported that a plant implementing the process produces 40,000 tons
of ethylene per year.

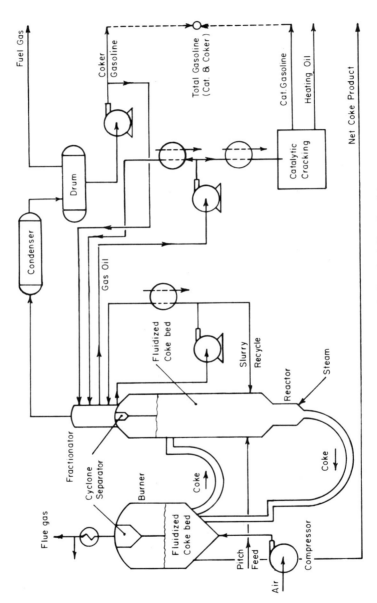

Fig. 16. Simplified diagram of the fluidized coking process.

2. Organic Chemicals

The production of phthalic anhydride via the solid catalyzed, exothermic oxidation of naphthalene is an extremely important example of the use of fluidized beds in the chemical process industries [36]. The primary problem associated with the process is the danger of explosion if the naphthalene concentration is too high. In order to prevent a reactor runaway, continuous cooling and close temperature control is necessary. Thus, to improve heat removal, the reactor is run as a fluidized bed with the incoming air fluidizing the solid catalyst.

The production of acrilonitrile (an important intermediate in nylon manufacture) by the Sohio process [37] is another example of a strongly exothermic solid-catalyzed oxidation reaction in which close temperature control is necessary for safe operation. In this process, propylene and ammonia are oxidized to produce acrilonitrile as well as hydrogen cyanide and acetonitrile.

A fluidized-bed process to remove the free phosphoric acid contained in the important fertilizer compound superphosphate has recently been reported [38]. The presence of free acids can cause corrosion in the production and handling equipment and adversely affect important physical properties of fertilizers. The free acids are removed by neutralization with ammonia; the authors report a new ammoniation process which takes advantage of a fluidized bed's large specific capacity and simple equipment construction.

As a final example, the solid-catalyzed, exothermic reaction of acetic acid and acetylene to produce vinyl acetate is carried out in a fluidized bed [4]. The vinyl acetate is an important raw material for the production of many polymer products.

3. Calcination

White and Kinsalla [39] report the fluidized-bed calcination of particles of limestone and dolomite by direct burning of a fuel in the bed itself. Since the reaction is very endothermic, multistaging is used to recover much of the necessary heat without large fuel consumption. As seen in Fig. 17 [4] the reactor has three stages: fuel

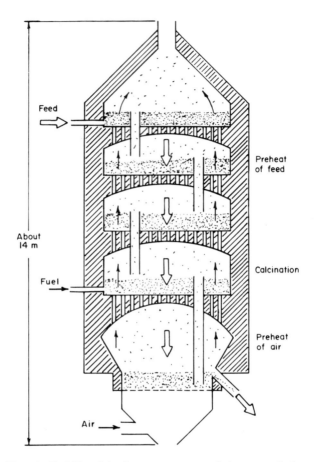

Fig. 17. A fluidized-bed process to calcine particles of limestone and dolomite.

enters the calcination stage and is burned in the presence of the solid phase with the fluidizing air; the air enters at the bottom stage; and the solid feed enters at the top stage. The air rising from the calcination stage preheats the falling feed in the top stage while the solid phase leaving the calcination stage preheats the rising air in the bottom stage.

Priestly [40] reports applications of fluidization to various calcining processes. For example, he discusses the calcining of phosphate rock in fluidized beds. Hydrocarbons may constitute 3.5% of a phosphate bed and this must be removed during the calcination process. Luckily, hydrocarbon content of most unweathered western United States phosphate ore furnishes most of the fuel required for calcination in a fluidized-bed reactor. Secondly, he reports that the Minerals, Pigments, and Metals Division of the Chas. Pfizer and Co., Inc., at its plant in North Adams, Mass., has replaced its conventional calciners with three fluosolids limestone calciners. This process is shown in Fig. 18. Finally, Priestly reports that the S. D. Warren Paper Company, at its mill in Muskegon, Mich., is using a fluidized-bed process for calcining pulp-mill lime mud; this represents a significant advance in recausticizing spent pulp-mill liquor.

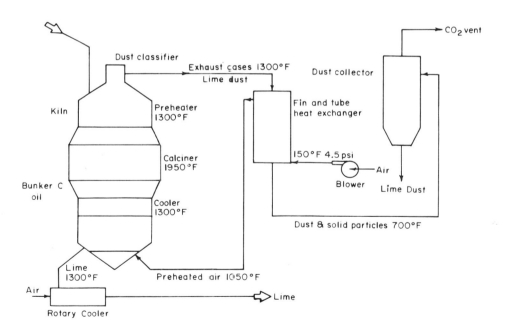

Fig. 18. Limestone calcination system.

4. Metallurgy

Tomasicchio [41] reports an unusual fluid-bed process that magnetically reduces iron ore. In this process, weakly magnetic, oxidized
iron compounds are converted by gaseous reducing agents to less-oxi-
dized iron compounds that are strongly magnetic. The high-grade con-
centrates are obtained by the magnetic separation of the iron compounds from the gangue. Dorr-Oliver has developed a process which
accomplishes the reduction in a fluidized bed; Tomasicchio reports
that a plant has been built by Dorr-Oliver and the Montecatini Edi-
son Company and is operating at Follonica, Italy. A schematic process flow sheet is shown in Fig. 19.

A Dorr-Oliver fluosolids fluidized process for the roasting of
sulfide ores has also been reported by Levenspiel and Kunii [4]. The
fluidized-bed design allows for higher capacity; the exothermic oxidation requires less excess air than alternative designs and thereby
an off gas richer in sulfur dioxide results.

5. Waste Treatment and Waste Removal

Bailie and Ishida [42] report a successful pilot-plant test of a
fluidized pyrolysis process for converting municipal refuse to gaseous
fuel. The process consists of two interconnected fluid beds; in one,
the organic waste is pyrolyzed while in the second, the char produced
from the pyrolysis reactions is combusted with air. The heat required by the pyrolysis reaction is supplied by circulating sand between the two fluidized beds. The Stamford Research Institute has
made a technical and economic analysis of the process [43].

Priestly reports a successful fluid-bed process for treating
waste liquors at the Green Bay Packaging Co.'s Green Bay (Wis.) sulfite pulp mill [40]. Spent sulfite liquor is fed to a fluidized-bed
combustion unit where water is evaporated and organics are oxidized.
The result is a high-quality sodium sulfate product. The liquor
combustion system is shown in Figure 20.

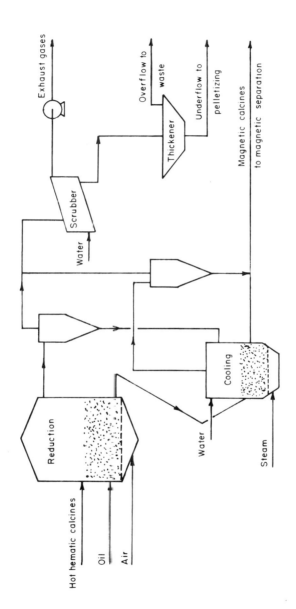

Fig. 19. Fluosolids reduction using direct fuel injection (DFI); installation at Montecatini's factory in Follonica, Italy.

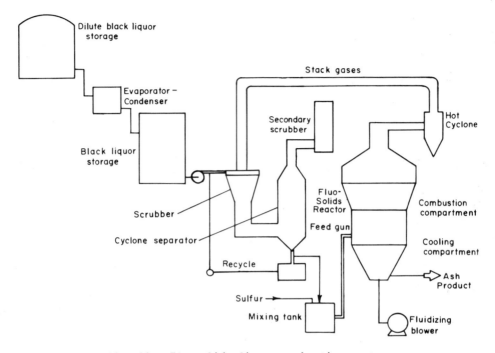

Fig. 20.   Fluosolids liquor combustion system.

6.   Food Processing

De Groot reports a continuous fluid-bed deep freezer for the production of frozen peas [44]. The freezer is designed to freeze 1,000 kg/hr of peas to -20°C with air at -30°C. The fluidized-bed freezer has been in operation successfully for a great variety of foods since 1961.

7.   Coal Gasification

Most processes for producing gas from coal are attempts to produce synthetic natural gas (SNG), a high-Btu substitute for natural gas (mostly methane). As reported by Conn [45], one of the four processes in an advanced stage of development (the so-called "hygas" process) is one in which hydrogen is contacted counter-currently with

coal in fluidized beds at high temperatures and pressures to produce
a gas rich in methane. As pointed out by Conn, demonstration on a
commercial basis is several years away although 75 ton/day pilot
plants have been put in operation.

The production of low-Btu gas from coal for power plants has become important because allowable sulfur dioxide emissions from power
plants have recently been reduced. Since the gas is produced under
reducing conditions, sulfur is converted to hydrogen sulfide (instead
of sulfur dioxide) which may be removed by well known absorption
processes. Thus, coal gasification presents an alternative to converting to a more expensive fuel, such as low-sulfur coals, or installing expensive facilities for stack gas scrubbing. Since the
only process to be commercialized (the Lurgi process) requires multiple units for large plants, interest in fluidized processing has been
intense. Consequently, a number of processes that gasify coal in
fluidized beds are being developed to take advantage of the large
capacities usually associated with fluid processing.

The Bamag-Winkler process [46] pictured in Fig. 21 [45], is an
example of a fluidized coal gasification process which has already
been developed [45]. Finely ground coal is fed to the gasifyer
where it meets an air-stream gas mixture flowing so that the coal is
maintained in a fluidized phase in the gasification zone. The ash
is removed at the bottom, the gas at the top; entrained ash and hydrogen sulfide are removed from the gas product in the gas clean-up
system.

Conn also reports that fluid-bed processes are in the development
stage [45] at Westinghouse and Union Carbide. The Union Carbide
gasification process, shown in Fig. 22 [45], is especially interesting because the heat needed in the gasifyer is transferred from the
combustor by continuously recirculating hot ash agglomerate between
the gasifyier and combustor. This is quite similar to the transfer
of heat via recirculating catalyst between regenerator and reactor
in the FCC process. Pilot-plant units are being built for both the
Westinghouse and Carbide processes.

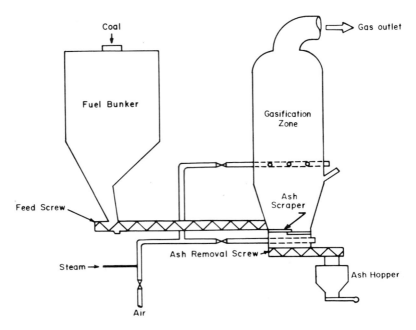

Fig. 21. An example of fluidized coal gasification.

8. Physical Processing

The applications of fluidized beds to such physical operations as mixing of fine powders, cooling or heating of fine solids, drying and sizing of powdery materials, and drying of gases via absorption of moisture on solids are discussed by Levenspiel [4]. Interested readers are referred to this discussion.

The process described for drying air by circulation of large silica gel beads is especially interesting because it bears some resemblance to the principles involved in the FCC process. The process is shown in Fig. 23. It is seen that the silica gel is recirculated from the drying column, where moisture present in the inlet air stream is removed by adsorption on the silica gel, to the reactivating column, where hot gases remove the moisture. This is similar to the regeneration and recirculation of cracking catalysts in the FCC process.

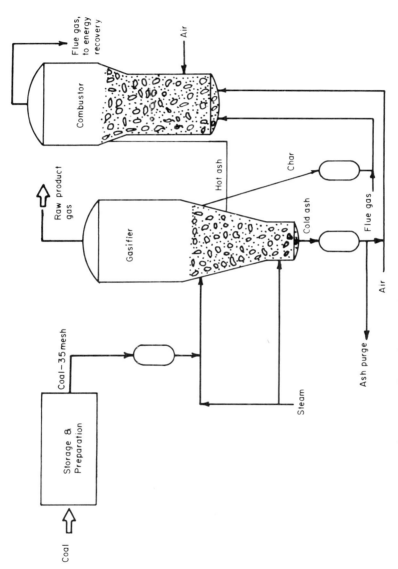

Fig. 22. The Union Carbide coal gasification process.

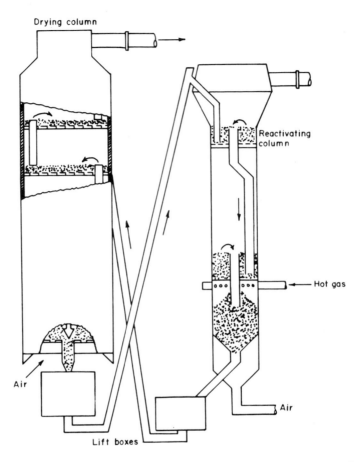

Fig. 23. Air drying via fluidized circulation of silica gel beads.

III.   SPOUTED BEDS

In a spouted bed gas enters the bed at the center of a conical bottom and is funneled through the center of the vessel. Only limited action is imparted to solids. Whether or not a bed of particles can be made to spout depends on such parameters as gas flow, bed depth, inlet-nozzle diameter, particle diameter, and column diameter. Figure 24 shows idealized phase diagrams that depict possible be-

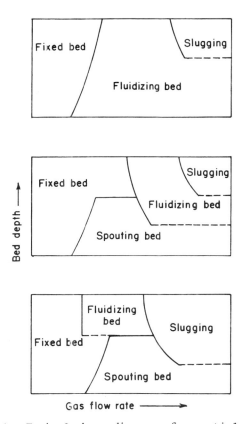

Fig. 24. Typical phase diagrams for particle beds.

havior of bed systems [2]. These idealized diagrams are similar to
the actual phase diagrams prepared from data collected by Mathur and
Gishler (47) on Ottawa sand of 20-35 mesh size. It is seen that at
some bed depths, increasing gas flow rates does not cause spouting;
the transition from fixed bed to fluidized bed occurs instead; and
at still larger gas velocities, slugging is encountered. As
discussed by Mathur and Gishler, the actual behavior depends on a
combination of the basic parameters mentioned. For this reason, experimental attempts have been made to empirically relate these variables. The results are summarized in the next section.

A. BASIC PARAMETERS

1. Minimum Fluid Velocity

For both water and air spouting, the minimum fluid velocity, U_{om}, that will maintain a spout is empirically expressed as [47]:

$$U_{om} = \left(\frac{d}{D_c}\right)\left(\frac{D_i}{D_c}\right)^b \left[2g L \frac{(\rho_p - \rho)}{\rho}\right]^{1/2}$$

where d = particle diameter, D_c = column diameter, D_i = fluid inlet diameter, g = gravitational constant, L = particle column depth, ρ_p = particle density, and ρ = fluid density. The exponent b is related to the shape of the column and has been determined for some designs [48].

2. Maximum Bed Height

The maximum spoutable bed depth L_m is given by Becker [49]:

$$L_m = 42 \left(\frac{d_e^{2 \cdot 76}}{D_c^{2 \cdot 76}}\right)\left(\frac{2.2 D_i}{D_c}\right)^n \left(22 + \frac{2,600}{Re_m}\right)^{2/3} \omega^{2/3} Re_m^{1/3}$$

where d_e is the equivalent spherical diameter, ω is a shape factor (= 1.0 for a sphere), n = 1.5 exp(- 0.0072Re_m), and Re_m is the modified particle Reynolds number.

3. Diameter of Spouting Section

Malek, Madonna, and Benjamin [50] present the following relationship for the proper diameter D_s for the spouting section of a bed:

$$D_s = (0.115 \log D_c - 0.031)G^{0.5}$$

where D_s is measured in inches, G = mass flow rate (lb/hr ft^2), and D_c is measured in inches.

4. Pressure Losses

    A typical pressure drop-flow diagram for a spouting bed [51] is
shown in Fig. 25.   The following regions may be identified:

    Branch a → b: Bed is fixed.

    Branch b → c: Spouting begins for velocities beyond point b.
    With further increases, the spout height increases, the bed depth
    increases, and pressure drop decreases.

    Branch c → d: No further increase in pressure; spout has not
    "broken through" as yet.

    Branch d → e: Spout penetrates through the bed boundary (incip-
    ient spouting), the concentration of solids in spout decreases
    abruptly, and pressure drop is reduced further.

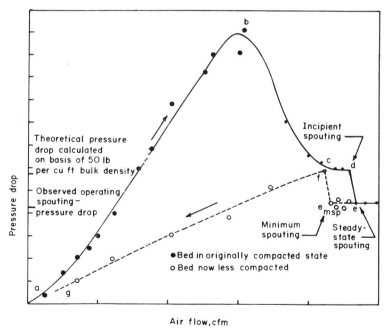

Fig. 25.   Typical pressure drop-flow diagram for a spouting bed.

Branch e and beyond: Steady-state spouting; i.e., pressure drop independent of mass velocity.

When the mass velocity is decreased, the pressure drop holds nearly constant until point e_{mps} is reached. For a further decrease in velocity, the spout submerges in the bed and there is a sudden increase in pressure drop; the point e_{msp} is the bed's minimum spouting condition. Further reduction of velocity from f to g is characteristic of fixed beds and thus, at f, the spout has completely collapsed.

B. GAS AND SOLID FLOW PATTERNS

Consideration of data on gas distribution between spout and annulus led Epstein and Mathur [52] to postulate an axial-dispersed plug-flow model for the gas flows in both the spout and the annulus. For the high particle Reynolds numbers encountered in both channels, axial mixing in either channel should be small [53].

Becker and Sallans [54], and Kugo, et al. [55] report experimental data from which they conclude that solids mixing in a spouted bed is nearly perfect. Ratcliffe and co-workers [56] provide further verification by invoking the mixed-flow models of Cholette and Cloutier [57] and Levenspiel [58] but ultimately conclude [59] that the perfect mixing assumption fits their data as well as the more complicated models first proposed. In addition, Chatterjee [60] proposes a model which describes the spouted bed in terms of two perfectly mixed tanks connected in series with the output of the second tank recycled to the first tank. Finally, Mann and Crosby [61] propose models which include the holdup in the spout, short circuiting, and flow in the bed proper.

C. HEAT AND MASS TRANSFER

Epstein and Mathur [52] present a comprehensive review of the literature on heat and mass transfer in spouting beds. The following paragraphs summarize the design equations discussed; interested readers are referred to this excellent article for complete discussions.

1. Heat Transfer Between Fluid and Particles

For 12-15 cm deep beds of several materials (d = 1-4 mm, ρ_p = 0.93-2.54 g/cm^3 spouted with air at 70°C in a 9.2-cm column, Uemaki and Kugo [62] present the following empirical equation:

$$\frac{hd}{k} = 0.005 \left(\frac{d_p U_{os} \rho_g}{\mu_g}\right)^{1.46} \left(\frac{U_o}{U_{os}}\right)^{1.30} \quad \text{(Prandtl number = 0.7)}$$

where h = heat transfer coefficient, d = particle diameter, k = thermal conductivity of particle, U_{os} = minimum superifical gas velocity for spouting, ρ_g = gas density, μ_g = gas viscosity, and U_o = superficial gas spouting velocity. Epstein and Mathur discuss why the preceding result should be considered only an approximation, although it does give some indication of the effect of important parameters on heat transfer.

2. Mass Transfer Between Fluid and Particles

For narrowly sized silica gel fractions of bed depth from 9.9-4.5 column diameters, spouted by air at 50°C in both an 8-cm and 10-cm diameter column, Uemaki and Kugo [63] present the following empirical equation:

$$\frac{k'D_c}{D_f} = 0.00022 \left(\frac{dU_o\rho_g}{\mu_g}\right)^{1.45} \left(\frac{D_f}{H_b}\right) \quad \text{(Schmidt number = 0.6)}$$

where k' = mass transfer coefficient, D_f = diffusivity of vapors in gas, D_c = column diameter, and H_b = bed depth. Again, Epstein and Mathur discuss the validity of the preceding results and the authors' methods.

3. Heat Transfer Between Wall and Bed

Empirical equations for gas spouting, based on dimensional analysis, have been proposed by Malek and Lu [64]:

$$\frac{h_w d}{k} = 0.54 \left(\frac{d}{H_b}\right)^{0.17} \left(\frac{d^3 \rho_g g}{\mu_g^2}\right)^{0.52} \left(\frac{\rho_p C_{ps}}{\rho_g C_{pg}}\right)^{0.45} \left(\frac{\rho_g}{\rho_p}\right)^{0.08}$$

where h_w = surface mean coefficient of heat transfer between wall and bed, C_{ps} = heat capacity of solid, and C_{pg} = heat capacity of gas, and Uemaki and Kugo [62]:

$$\frac{h_w d}{k} = 13.0 \left(\frac{d_o}{d}\right)^{0.2} \left(\frac{dU_o \rho_g}{\mu_g}\right)^{0.10} \left(\frac{d^3 \rho_g g}{\mu_g^2}\right)^{0.46} \left(\frac{\rho_p C_{ps}}{\rho_g C_{pg}}\right)^{-0.42} (1 - \varepsilon)$$

where d_o = inlet orifice diameter and ε = average spouted bed voidage.

Epstein and Mathur point out the serious contradictions between the two results and, in general, the inadequacy of these empirical correlations. Instead, Epstein and Mathur propose a more rational approach, based on the concept of a thermal boundary layer, to correlate h_w using the penetration model concept. Their result is:

$$h_w = 1.129 \left(\frac{U_w \rho_b C_{ps} k_{eb}}{H_b}\right)^{1/2}$$

where U_w = particle velocity at the wall, ρ_b = bulk density of the bed, k_{eb} = effective thermal conductivity of the bed, and H_b = heated length. Figure 26 compares their estimated values of h_w with measured values, and it is seen that even though the actual data are clustered in a small region, the order-of-magnitude of the theoretical prediction is correct. It is also reported that the qualitative trends in the observed results are substantiated by their model.

D. CHEMICAL REACTIONS AND INDUSTRIAL APPLICATIONS

Uemaki et al. [65,66] have recently demonstrated for the thermal cracking of petroleum that spouted beds may be more favorable than other fluid-solid systems for carrying out certain types of reactions.

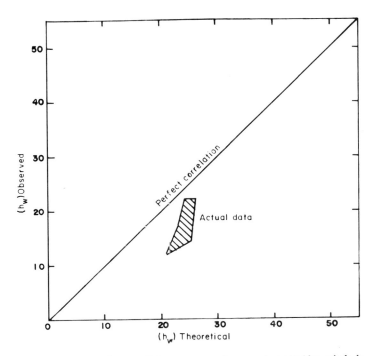

Fig. 26.   Comparison of h_w measured experimentally with h_w predicted by penetration model.

Mathur and Lim [67] have developed a theoretical model of a spouted-bed chemical reactor involving vapor-phase reaction in the presence of a solid catalyst.   The application of the model showed that the possibility of achieving better conversion by using a spouted bed instead of a fluidized bed is confined to relatively fast reactions for which the reaction rate is independent (or weakly dependent) on catalyst particle size.

An early application of spouted-bed technology is the pilot-scale wheat drying process [68] pictured in Fig. 27 [51]. This process is discussed in the chapter on spouted beds in Leva's textbook [1].

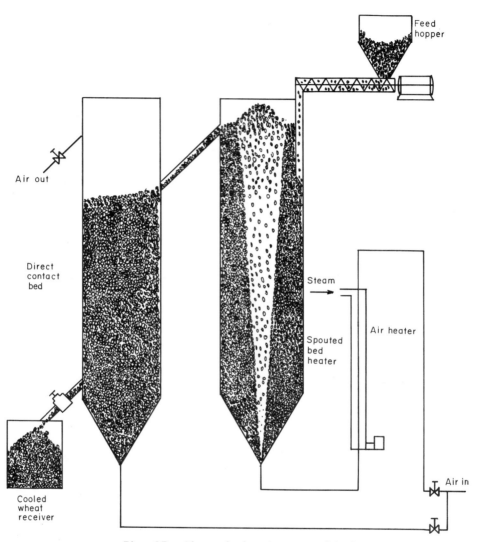

Fig. 27. Wheat drying in spouted bed.

Many more applications are discussed by Mathur in his chapter on
spouted beds which appears in Davidson and Harrison's book on fluidization [24]. Table 3 is taken from Mathur's chapter and illustrates
the variety of process applications reported in the literature; fur-
thur details may be obtained from reference 24.

TABLE 4

Applications of Spouted Beds

Application	Particulars
1. Cooling fertilizers	Double-deck installation with multiple spouts for cooling up to 30 tons/hr of fertilizer from 120-40°C.
2. Preheating coal	Coal of about 6 mm size heated to 250°C in continuous operation, as pretreatment before coking in coke ovens. Promising outcome, multistage operation visualized for large scale.
3. Low-temperature coal carbonization	Use of coarse particles (2.5 mm) together with the violent agitation in the spout region allowed continuous operation without the problem of agglomeration. Process worked well with a variety of Australian coals at carbonization temperatures 450-650°C.
4. Pyrolysis of shale	Continuous operation with coarse shale up to 6 mm in size, temperatures of 950 to 1,350°F. Attrition of particles in the spout region was beneficial, since the outer surface of a particle became fragile on loss of organic matter and was broken off, exposing fresh surface for retorting. Fine spent shale collected in an overhead cyclone.
5. Tablet coating	Batch operation. Coating solution atomized into the bed through a small nozzle located at the center of the hot-air orifice. Compared to conventional coating pans, spouted bed gave more uniform coating, better batch-to-batch uniformity, shorter batch time, and lower overall cost.
6. Granulation	Initial bed consists of nuclei of the product granules. Melt or solution of the material is atomized into the bed as in the coating process, but with continuous discharge of the finished granules. Essentially the same method as above applied to granulating a variety of substances in the form of solutions, suspensions, and pastes. Process combines fertilizer granulation with neutralization in the same spouted-bed unit. Acid spray and ammonia vapors are introduced into the spouting air below the inlet orifice. Bed consists of product granules.
7. Solids blending	Rapid and effective blending of granular solids achieved at lower power cost than in mechanical belnders. A vertical pipe with openable ports introduced in the axial region to economize on air pressure required to initiate spouting.

NOMENCLATURE

A　　Cross-sectional area

b　　Exponent, related to column shape, needed to calculate U_{om}

B　　Factor in equation

C_d'　Orifice constant

C_R　Correction factor for nonaxial location of immersed tubes

C_{ps}　Heat capacity of solid

C_{pg}　Heat capacity of gas

d_p　Particle diameter

d_e　Equivalent spherical diameter

d_t　Diameter of tube

d_o　Inlet orifice diameter

d_{or}　Orifice diameter

E　　Fluidization parameter

D_c　Column diameter

D_i　Fluid inlet diameter

D_s　Spouting-bed diameter

D_f　Diffusivity of vapors in gas

Fr　　Froede dimensionless number

g　　Gravitational acceleration

g_c　Gravitational constant

G　　Mass flow rate

G'　　Mass flow velocity

h　　Local heat transfer coefficient

H_b　Bed height

k	Thermal conductivity
k_g	Thermal conductivity of gas
k'	Mass transfer coefficient
k_{eb}	Effective thermal conductivity of the bed
L	Particle column depth
L_f	Fluidized bed depth
L_m	Maximum spoutable bed depth
L_{mf}	Minimum fluidization bed depth
ΔP	Pressure drop of fluidized bed
ΔP_d	Pressure drop of distributor
ΔP_T	Total pressure drop
q	Rate of heat transfer
Q	Volumetric gas rate
N_{or}	Number of orifices
R	Fluidization parameter
Re	Reynolds number
Re_m	Modified particle Reynolds number
T	Temperature difference
U_o	Superficial velocity
U_w	Particle velocity at the wall
U_{om}	Minimum superficial fluid velocity for spouting
U_{mf}	Minimum superficial fluidization velocity
W	Weight of particles
ε	Average bed porosity or void fraction
μ	Fluid viscosity

μ_g Gas viscosity

ρ Fluid density

ρ_p Particle density

ρ_g Gas density

ϕ Sphericity of solid particles

ω Shape factor necessary to calculate L_m

REFERENCES

1. M. Leva, Fluidization, McGraw-Hill, New York, 1959.

2. C. Orr Jr., Particulate Technology, McMillan, New York, 1966.

3. C.Y. Wen and Y.H. Yu, A.I.Ch.E. J., 12, 610 (1966).

4. D. Kunii and O. Levenspiel, Fluidization Engineering, Wiley, New York, 1969.

5. P.H. Pinchbeck and F. Popper, Chem. Eng. Sci., 6, 57 (1956).

6. W.J. Rice and R.H. Wilhelm, A.I.Ch.E. J., 4, 423 (1958).

7. J.B. Romero and L.N. Johanson, Chem. Eng. Prog. Symp. Series, 58, (38), 28 (1962).

8. N.J. Hassett, "A Critique of the Two-Phase Theory of Bubbling Fluidization", Proceedings of the International Symposium on Fluidization, Netherlands University Press, Amsterdam, 1967, p. 362.

9. E.W. Grohse, A.I.Ch.E. J., 1, 361 (1955).

10. W.W. Shuster and P. Kisliak, Chem. Eng. Prog., 48, 455 (1952).

11. F.A. Zenz and D.F. Othmer, Fluidization and Fluid-Particle Systems, Reinhold, New York, 1960.

12. R.D. Toomey and H.F. Johnstone, Chem. Eng. Prog. Symp. Series, 49, (5), 51 (1953).

13. O. Levenspiel and J.S. Walton, Chem. Eng. Prog. Symp. Series, 50, (9), 1 (1954).

14. W.M. Dow and M. Jakob, Chem. Eng. Prog., 47, 637 (1951).

15. G. Van Heerden, P. Nobel, and D.W. van Krevelen, Chem. Eng. Sci., 1, 51 (1951).

16. C.Y. Wen and M. Leva, A.I.Ch.E. J., 2, 482 (1956).

17. L. Wender and G.T. Cooper, A.I.Ch.E. J., 4, 15 (1958).

18. H.A. Vreedenberg, J. Appl. Chem. (London), 2, Suppl. 1, 526 (1952).

19. H.A. Vreedenberg, Chem. Eng. Sci., 9, 52 (1958).

20. H.A. Vreedenberg, Chem. Eng. Sci., 11, 274 (1960).

21. N.M. Levitz, E.R. Gilliland, and W.C. Bauer, Ind. Eng. Chem., 41, 1104 (1949).

22. M.F. Quinn, Ind. Eng. Chem., 55, 18-24 (1963).

23. R.J. Aldrich, "Fluid Bed Dryers," Encyclopedia of Process Equipment, Reinhold, New York, 1964.

24. J.F. Davidson and D. Harrison, Fluidization, Academic Press, New York, 1971.

25. J.F. Davidson and D. Harrison, Fluidized Particles, Cambridge University Press, Cambridge, England, 1963.

26. D.L. Pyle, Advances in Chemistry Series 109, Chemical Reaction Engineering, American Chemical Society, Washington, D.C., 1972.

27. K. Kato and C.Y. Wen, Chem. Eng. Sci., 24, 1351-1369 (1969).

28. A.B. Whitehead and A.D. Young, Proceedings of the International Symposium on Fluidization, Netherlands University Press, Amsterdam, 1967, p. 234.

29. J.H. de Groot, Proceedings of the International Symposium on Fluidization, Netherlands University Press, Amsterdam, 1967, p. 348.

30. A.L. Conn, Chem. Eng. Prog., 69, No. 12, 11-17 (1973).

31. D.F. Othmer, Ed., Fluidization, Reinhold, New York, 1956.

32. R.J. Hengstebeck, Petroleum Processing, Principles and Applications, McGraw-Hill, New York, 1959.

33. W.S. Kmak, paper presented at A.I.Ch.E. National Meeting, Houston, Texas, March 1971.

34. W.L. Nelson, Petroleum Refinery Engineering, McGraw-Hill, New York, 1958.

35. A. Steinhofer, Hydrocarbon Processing and Petroleum Refiner, 44, 134 (1965).

36. J.J. Graham and P.F. Way, Chem. Eng. Prog., 58, 96 (1962).

37. F. Veatch, Hydrocarbon Processing and Petroleum Refiner, 41, 187 (1962).

38. J. Trojar and V. Vanecek, Brit. Chem. Eng., 10, No. 11, 756-759 (1965).

39. F.S. White and E.L. Kinsalla, Mining Eng., 4, 903 (1952).

40. R.J. Priestly, "High Temperature Reactions in a Fluidized Bed," Proceedings of the International Symposium on Fluidization, Netherlands University Press, Amsterdam, 1967, pp. 701-710.

41. G. Tomasicchio, "Magnetic Reduction of Iron Ores by the Fluosolids System," Proceedings of the International Symposium on Fluidization, Netherlands University Press, Amsterdam, 1967, pp. 725-735.

42. R.C. Bailie and M. Ishida, A.I.Ch.E. Symposium Series, 68, No. 122 (1972).

43. R.C. Bailie and S.B. Alpert, "Use of Fluidized Beds for Conversion of Municipal Refuse to Fuel Gas," to be published.

44. W. Herman de Groot, "Continuous Fluid-Bed Deep Freezer for Peas," Proceedings of the International Symposium on Fluidization, Netherlands University Press, Amsterdam, 1967, pp. 566-572.

45. A.L. Conn, Chem. Eng. Prog., 69, No. 12, 56-61 (1973).

46. Bituminous Coal Research, Inc., "Gas Generator Research and Development, Survey and Evaluation," Phase 1, Vol. 2. Report prepared for Office of Coal Research, U.S. Department of the Interior, under contract No. 14-01-0001-324, March 1965, pp. 212-216.

47. K.B. Mathur and P.E. Gishler, A.I.Ch.E. J., 1, 157-164 (1955).

48. L.A. Madonna, R.F. Lama, and W.L. Brisson, Brit. Chem. Eng., 6, 524-528, (1961).

49. H.A. Becker, Chem. Eng. Sci., 13, 245-262 (1961).

50. M.A. Malek, L.A. Madonna, and C.Y. Benjamin, Ind. Eng. Chem., 2, 30-34 (1963).

51. B. Thorley, K.B. Mathur, J. Klassen, and P.E. Gishler: "Report on Effect of Design Variables on Flow Characteristics in a Spouted Bed", National Research Council, Ottawa, Canada (1955).

52. N. Epstein and K.B. Mathur, Can. J. Chem. Eng., 49, 467-476 (1971).

53. N. Epstein, Can. J. Chem. Eng., 36, 210 (1958).

54. H.A. Becker and H.R. Sallans, Chem. Eng. Sci., 13, 97 (1961).

55. M. Kugo, N. Watanabe, O. Uemaki, and T. Shibata, Bull. Hokkaido Univ. Japan, 39, 95 (1965).

56. R.K. Barton, G.R. Rigby, and J.S. Ratcliffe, Mech. Chem. Eng. Trans., Australia, 4, 105 (1968).

57. A. Cholette and L. Cloutier, Can. J. Chem. Eng., 37, 105 (1959).

58. O. Levenspiel, Can. J. Chem. Eng. 40, 135 (1962).

59. M.J. Quinlan and J.S. Ratcliffe, Mech. Chem. Eng. Trans., Australia, 6, 19 (1970).

60. A. Chatterjee, Ind. Eng. Chem. Process Design Develop., 9, (4) 531-536 (1970).

61. U. Mann and E.J. Crosby, Ind. Eng. Chem. Process Design Develop., 11, (2) 314-316 (1972).

62. O. Uemaki and M. Kugo, Kagaku Kogaku, 31, 348 (1967).

63. O. Uemaki and M. Kugo, Kagaku Kogaku, 32, 895 (1968).

64. M.A. Malek and B.C.Y. Lu, Can. J. Chem. Eng., 42, 14 (1964).

65. O. Uemaki, M. Fugikawa, and M. Kugo, Kogyo Kagaku Zasshi, 73, 453 (1970).

66. O. Uemaki, M. Fugikawa, and M. Kugo, Kogyo Kagaku Zasshi, 74, 933 (1971).

67. K.B. Mathur and C.C. Lim, Chem. Eng. Sci., 29, 789-797 (1974).

68. K.B. Mathur and P.E. Gishler, J. Appl. Chem. (London), 5, 624 (1955).

H. William Blakeslee

Industrial Emissions Department
Scott Environmental Technology, Inc.
Plumsteadville, Pennsylvania

I.   INTRODUCTION

Powder metallurgy encompasses a set of processes in which finely
divided metal powders are converted into coherent objects. It is
usually accomplished through the use of pressure (compaction) to consolidate the powder into the desired shape followed by heat (sintering) to fuse the particles together. Various alternative procedures,
such as heat and pressure applied simultaneously as in hot isostatic
compaction or heat (sintering of lightly compacted preforms) followed

195

by pressure (forging) to produce high-strength parts, are useful for special purposes. Sintering without prior compaction is used to produce objects with low density and also in a process called slip casting. Pressure without heat can produce coherent objects where either the pressure is high enough or the rate of its application is extremely high. The factor common to all these variations of the powder metallurgy process is that at no time is the temperature high enough to fuse the major portion of the metallic matrix. This distinguishes powder metallurgy from the rest of metallurgy.

The growth of the powder metallurgy industry is based primarily on economic factors. Certain types of close-tolerance metal objects can be manufactured at high production rates in automatic equipment with no scrap. These parts are generally equivalent to machined parts and involve far less labor. They are stronger than cast parts and can be made to higher dimensional tolerances. Figure 1 shows typical powder metallurgy parts.

Increasing in importance are the facets of powder metallurgy that utilize its unique properties, which can be produced by no other technology. Foremost among these is the ability to control porosity in the finished part from less than 1% to over 60%. The earliest use of porosity control was in the production of oilless bearings, which are now manufactured at a rate of more than 3 billion per year. More recently, the use of porous sintered preforms for forging has shown the potential of revolutionizing the forging industry if and when the price of metal powders becomes more competitive with wrought metal stock.

Other unique properties of powder metallurgy include the fabrication of parts from metals which cannot be easily fused such as tungsten, tantalum, niobium, and beryllium, and the formation of composites and alloys of materials that cannot be melted together for various reasons. Examples of the latter include dispersion-strengthened metals, copper infiltrated steels, and some high-temperature superalloys. The powder metallurgy process extends the degree to which the metallurgist can vary the matrix of a metallic object by a very large degree.

Fig. 1. Typical parts made by powder metallurgy. (Courtesy of the Sharples - Stokes Division of the Pennwalt Corporation.)

The gas-solid interface enters into most phases of the powder metallurgy process. It is controlled in the production of the metal powders in order to reduce the amount of oxide and other impurities and control the particle size and compaction properties; it must be taken into consideration in the storage and handling of powders prior to compaction due to the ease of contamination and also because of the hazards inherent in fine metal powders; it must be controlled during compaction often by vacuum or by inert gases; and finally, it is the most important variable during sintering.

II. POWDER PREPARATION

Gas-solid interactions are of primary importance in the production
and handling of metal powders because of the high surface-to-volume
ratios involved. Two of the more common methods of powder production
utilize primary gas-solid reactions. These are the reduction of metal
oxides with hydrogen or carbon and the decomposition of metal carbonyls.
Other powder production methods such as atomization, electrolytic
deposition, and comminution also involve controlled reactions of the
metal powders with the surrounding gas phase to produce certain required properties. Tables 1 and 2 summarize the properties of the
most common methods of powder production. In most cases, the gas-
solid reactions occurring during powder production will effect the
properties of the compacted and sintered final parts.

The reduction of various forms of iron oxide is the most common
commercial powder production method at this time. This is due to the
low cost of the raw material, the economies of the process, and the
applicability of the final product in most press and sinter operations.
In addition to iron, copper, nickel, cobalt, tungsten, and molybdenum
are also produced in powder form by this method. The resulting irregularly shaped particles are from 98.5 to 99% pure and compact easily
in standard presses.

The commercial reduction of iron ore is known as the Höganäs process and is achieved by layering powdered magnetite ore with coke and
limestone. The mixture is heated to 1,200°C and produces a sinter
cake sometimes known as sponge iron. It is ground up, magnetically
separated, and annealed and purified in a hydrogen atmosphere furnace.
Some of the reactions occurring in the kiln are:

$$CaCO_3 \text{ (solid)} \longrightarrow CaO \text{ (solid)} + CO_2 \text{ (gas)}$$
$$CO_2 \text{ (gas)} + C \text{ (solid)} \longrightarrow 2CO \text{ (gas)}$$
$$4CO \text{ (gas)} + Fe_3O_4 \text{ (solid)} \longrightarrow 3Fe \text{ (solid)} + 4CO_2 \text{ (gas)}$$

A similiar process involving hydrogen reduction is used to produce
iron powders from mill scale and iron scrap. These raw materials are

ground to a powder and uniformly oxidized in air at 870 to 980°C. They
are then reduced to the metal with hydrogen gas at 980°C. An iron
sinter cake results which is pulverized by crushing or milling.

High purity, fine powders of iron, nickel, and cobalt can be produced by the carbonyl process. This utilizes the high volatility of
certain metal carbonyls to separate the metal from the impurities.
The raw materials can be low grades of metal powder which are reacted
with carbon monoxide at high pressure and temperature to form a gaseous
carbonyl of the metal. This is vented from the reactor and condensed.
Since it is a liquid under ambient conditions [b.p.$Ni(CO)_4$ = 43°C,
$Fe(CO)_5$ = 107°C], it can be stored in tanks. Conversion to the metal
powder is accomplished by spraying the liquid carbonyl into a heated
chamber which decomposes the carbonyl. The reversible reaction in the
process is:

$$xM \text{ (solid)} + yCO \text{ (gas)} \rightleftharpoons M_x(CO)_y \text{ (gas)}$$

Iron pentacarbonyl is produced at 250 to 300°F and at 150 to 200
atm and is decomposed at about 460°F. The iron particles form in the
decomposition chamber in a spherical shape by alternating layers of
carbon and iron. The carbon results from the decomposition of carbon
monoxide to form carbon dioxide and carbon. Some nitrogen (0.8%) will
also contaminate the powder if it is present in the chamber atmosphere.
The powder can be used in this form or decarburized in hydrogen at
800 to 1,000°F. This removes the layered structure of the powder and
eliminates both the carbon and the nitrogen impurities. A spherical
powder of 99.5% purity in the low micrometer-size range results.

The decomposition of the metal carbonyls can be considered a precipitation from the vapor phase. Other examples include the condensation of zinc vapor and the reduction of metal halide vapors such as
titanium tetrachoride by metallic magnesium.

A common method of producing titanium and zirconium powder is
through the hydride-dehydride process [2]. This depends on the great
solubility of hydrogen in these metals and their alloys at elevated

TABLE 1

Comparison of Principal Commercial Methods of Producing Metal Powders[a]

Method	Raw materials	Powders produced	Advantages	Disadvantages	Relative cost
Atomization	Scrap or virgin melting stock or metal or alloy powder desired	Stainless steel, brass, bronze, other alloy powders, Al, Sn, Pb, Fe, Zn	Best method for alloy powders. Applicable to any metal or alloy melting below 3,000°F	Wide range of particle sizes, not all salable. Particles too spherical for some applications	Low to medium
Gaseous reduction of oxides	Oxides of metals such as Cu_2O, NiO, Fe_3O_4	Fe, Cu, Ni, Co, W, Mo	Easy to control particle size of powder. Good compacting powder	Requires high grade oxides. Restricted to reducible oxides	Low
Gaseous reduction of solutions	Ore for leaching or other metal salt solution	Ni, Co, Cu	Ore can be used. Purification during leaching. Fine particles	Applicable to few metals such as Ni, Co, Cu	Medium

200

Method	Raw material	Metals	Advantages	Limitations	Cost
Reduction with carbon	Ore or mill scale	Fe	Low cost. Control of particle size, controlled variation in properties possible	Requires high grade ore or mill scale. Applicable mainly to iron	Low
Electrolytic	Generally soluble anodes of iron and copper	Fe, Cu, Ni, Ag	High purity of product. Easy to control	Limited to few metals, cost	Medium
Carbonyl decomposition	Selected scrap, sponge, mattes	Fe, Ni, Co	Produces fine pure powders	Limited to few powders, high cost	High
Grinding	Brittle materials such as Be, high sulfur nickel, high carbon iron, Sb, Bi, Fe, Mn cathodes	Fe, Be, Mn, Ni, Sb, Bi	Controlled size of powder	Limited to brittle or embrittled materials. Quality of powder limits use. Slow	Medium

aReprinted from reference 1, p. 15 by courtesy of Plenum Press.

TABLE 2

Some Distinguishing Characteristics of Metal Powders Made by Various Commercial Methods[a]

Method of production	Typical purity[b] (est.)	Particle characteristics Shape	Meshes available	Compressibility (softness)	Apparent density	Green strength	Growth-with-copper-of iron[c]
Atomization	High 99.5+	Irregular to smooth, rounded dense particles	Coarse shot to to 325 mesh	Low to high	Generally high	Generally low	High
Gaseous reduction of oxides	Medium 98.5 to 99.+	Irregular, spongy	Usually 100 mesh and finer	Medium	Low to medium	High to medium	Low
Gaseous reduction of solutions	High 99.2 to 99.8	Irregular, spongy	Usually 100 mesh and finer	Medium	Low to medium	High	Iron not produced by this method
Reduction with carbon	Medium 98.5 to 99.+	Irregular, spongy	Most meshes from 8 down	Medium	Medium	Medium to high	Medium
Electrolytic	High+ 99.5+	Irregular, flaky to dense	All mesh sizes	High	Medium to high	Medium	High
Carbonyl decomposition	High 99.5+	Spherical	Usually in low micrometer ranges	Medium	Medium to high	Low	?
Grinding	Medium 99.+	Flaky and dense	All mesh sizes	Medium	Medium to low	Low	High

[a]Reprinted from reference 1, p. 16 by courtesy of Plenum Press. [b]Purity varies with metal powder involved. [c]Growth-with-copper of iron during sintering is incresed in radial dimension of compacted iron-plus-copper powders.

temperatures. The hydrogen is not very soluble at room temperature
and precipitates out as a brittle hydride. This can be easily pulverized and the hydrogen removed to yield a rough, equiaxed powder
with good compaction properties. The process for titanium starts with
an ingot of the pure or alloyed metal which is heated in an evacuated
chamber to 840°F. At this temperature, hydrogen is introduced and the
chamber allowed to cool to room temperature converting the ingot to
titanium hydride. This is broken up, then crushed, and finally put
through an attrition mill to obtain the desired size distribution, all
under an atmosphere of argon. Dehydriding is performed under vacuum
at 1,290 to 1,380°F for as long as 24 hr. The resulting lightly sintered product is attrited and screened.

The atomization process yields fine particles by dispersing molten
metal with a stream of high-pressure fluid or by several other means.
It is a valuable method of producing high-purity powders of iron, aluminum, tin, lead, and zinc. In fact, it is increasingly replacing other
powder manufacturing methods at such a rate that in 1970, 60% of the
iron powder produced in the United States and Canada was made by atomization [3]. Its most important application is in the production of
steels and high-temperature alloys used in the newer PM (powder metallurgy) techniques such as forging of preforms and extrusion of 100%
dense mill shapes, and in the manufacture of tool steels. The advantages
of atomized powders are [3]:

1. Any alloy can be made including new alloys and systems designed
especially and solely for PM.

2. Same and uniform composition of all particles: macrosegregation
eliminated; more even distribution of constituents (eg., carbides);
grain-size homogeneity; improved workability and reproducibility of
properties; improved material yield.

3. Control of particle shape, size, and structure: spherical to
irregular or splat; homogeneous and fine-grain structure; ease of
metalworking; reproducibility of properties.

4. Higher purity: fewer nonmetallic inclusions than reduced
powders; high compressibility; improved properties and structures.

5. Lower capital costs: investment can be tailored to market
needs with greater flexibility.

Alloys have been traditionally produced in powder metallurgy by
mixing various pure metal powders in the desired ratios before compacting and sintering. Metals with widely differing melting points
or which react with the sintering atmospheres cannot be alloyed in
this manner, therefore, the types of alloys that could be used in
powder metallurgy were limited. The use of prealloyed powders made
by atomization removes this restriction. These, however, often consist of very hard particles and are generally expensive, so the use
of "master alloy" powders which can be subsequently mixed with various
proportions of pure metal powders before compacting is sometimes a
useful compromise.

Brass, bronze, and stainless steels are the most common alloy
powders made by atomization. Others include hard facing alloys, magnetic alloys, high-speed steels, and various cobalt - and nickel-based
superalloys. These powders are atomized in inert gas and are consequently spherical so they do not compact easily in conventional processes. Densification is often accomplished for the high-temperature
alloys by hot consolidation techniques such as sealing in mild steel
cans and extruding or rolling. For other alloys, the variations in
the atomization process can produce irregular particles which can be
compacted more easily. Also, the more brittle alloys can be ground
following atomization to produce powders which can be compacted either
alone or mixed with elemental powders.

The oldest commercial atomization process for iron consists of the
disintegration of a molten stream of high-carbon iron by high pressure
jets of water or air. In the Mannesmann process, a low-carbon steel
scrap is melted in a cupola or electric furnace and then poured through
a high-pressure stream of air which breaks the stream into fine droplets that are caught and quenched in a water tank. By controlling the
carbon in the melt, the decarburation in the air is not complete so

that when the dried powder is heated up to 1,250°F, the remaining carbon combines with the oxide layer to form carbon monoxide. A pure iron sinter cake results which is crushed, sieved, and blended.

Figure 2 shows a generalized schematic of most atomization processes and lists the variables which determine the size, shape, and chemistry of the final powder. The following data serve as an example of one set of parameters used to produce water-atomized, irregular stainless steel powder with 38% of the particles finer than 100 mesh [6]:

Metal temperature	1,510°C
Metal nozzle diameter	¼ in.
Metal flow rate	48 lb/min
Angle of metal stream to water stream	50° (8 nozzles)
Water pressure	1,300 psi

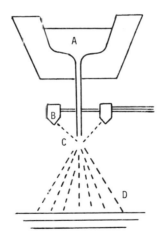

Fig. 2. Variables of the atomization process (after Ref. 3)
A: molten metal - atmosphere, chemistry of alloy, viscosity, surface tension, melting temperature range, superheat, metal feed rate, nozzle diameter; B: atomizing agent - gas or liquid, pressure, flow rate, velocity, viscosity; C: atomization - spread between jets, length of atomizing stream, length of metal stream, angle of atomizing stream to metal stream; C: atmosphere, flight path of particles, quenching medium.

Water flow rate	44 gal/min
Water velocity	370 ft/sec

The total surface free energy of a stream of molten metal is increased when it is dispersed into droplets. The increase is proportional to the surface tension of the metal, and inversely proportional to the droplet diameter. This additional energy is supplied by the kinetic energy of the dispersing medium. The design of the atomizing equipment must allow for the quenching of the newly formed particles as soon as possible in order to prevent coalescing of the fine droplets. The use of water as the dispersing medium accomplishes this and in addition freezes the droplets in their initial irregular shapes to yield powders with good green strength in conventional pressing. Impacting the molten particles onto a cooled metal surface, splat cooling, is another way of preserving an irregular shape.

Atomization by inert gases produces particles that assume a spherical shape almost immediately because of the high surface tension of the molten metals. Additions to the melt of magnesium in copper alloys [4] or silicon in steel [5] decrease the surface tension sufficiently to produce more irregular shapes. The decrease is generally due to the formation of an oxide layer onto the particles. The presence of oxides in alloy powders must be avoided since they are often refractory. Argon can be used as the disintegrating medium when the atomization is done in a sealed chamber. The molten metal is poured through the top and the argon recirculated to reduce production costs.

A variation of the atomization process has been developed for the production of pure titanium and titanium alloy powders. It is called the rotating electrode process [2]. An electrode made from the desired alloy mounted in a chuck is rotated at high speed in the center of an eight foot drum. An arc is struck between this electrode and a stationary one made of tungsten which causes the titanium to melt and disperse into fine drops which are hurled to the outside wall of the drum. The atmosphere in the drum is purified helium so the droplets harden with a minimum of impurities. The resulting powder can be hot pressed or extruded.

Metal powders in a wide range of particle shapes can be produced
by the electrolytic process. It is used on iron, copper, nickel, and
silver and produces powders of high (99.5%) purity which are easily
compacted. They are also expensive, however, so are only useful in
special applications. The process uses a soluble salt of the metal
which is electrolyzed under conditions favoring the formation of a
coarse mass on the cathode or a sludge, either of which can be easily
converted into a fine powder. The result is a dendritic powder which
has excellent compaction properties after annealing in a reducing
atmosphere.

There are various methods of reducing large particles of metal
into powders fine enough for powder metallurgy. These are all classified as milling processes and involve some form of impaction to reduce
particle size. Ball and vibratory mills agitate mixtures of powders
and refractory objects to produce comminution. This is normally done
in a controlled atmosphere, but may also be done in a liquid medium.
Other milling methods include the acceleration of particles by a high-
velocity gas stream and impacting them on a hard surface to reduce
their size.

The latter process is called the coldstream process. It utilizes
low-temperature brittleness to produce an iron powder with a very
small and uniform particle size, irregular shapes, and nearly oxide-
free surfaces. The process uses a high-velocity, high-pressure air
stream shooting metal powder through a venturi nozzle at a target in
an evacuated glass chamber. At the exit of the nozzle, the pressure
drops instantaneously from about 100 psi to atmospheric causing the
temperature to drop below zero. The chilled metal shatters when it
impacts on the target. A system of classifying operations is used to
separate the finished powders from the larger fractions which are
recycled.

Milling is most successful on brittle materials such as beryllium,
high-sulfur nickel, high-carbon steel, antimony, and bismuth. It
yields flakey and dense particles of medium (99%) purity and is usually
very slow. The particle-size distribution is a function of the material

being ground and the conditions. Chromium, for example, is brittle
down to a particle size roughly equal to its grain size and is ductile
below this size. The particle size obtainable by ball milling in inert
environment approximates the grain size of about 0.5 µm. In a reactive
atmosphere, however, particle sizes below 0.06 µm can be reached for
chromium powders. Such extremely fine powders have been used in research to obtain finely dispersed metal composites.

The limitation to particle-size reduction during ball milling is
the point at which the agglomeration rate of the particles equals the
comminution rate [7]. Agglomeration is caused by welding of the particles and is reduced in materials which become work-hardened by the
milling process. Surface-active agents used in wet milling reduce
agglomeration, but the most effective preventitive is the formation
of a reaction layer on the freshly exposed metal surfaces before they
contact another particle. This happens automatically with the more
reactive metals when they are milled in air. Metals with easily
reduced oxides can be ground to very fine powders in this manner, but
in metals with difficult-to-reduce oxides the powder may sinter because such high temperatures are required to reduce the oxides. These
metals can be ground in the presence of other reactive gases, however,
which produce products more easily reduced. Submicrometer nickel and
chromium powders have been made by ball milling in the presence of
hydrogen halide vapors [8].

Ball milling in water has produced extremely fine metal powders
because of oxidation reactions. Table 3 summarizes the results of a
research program by NASA in this area [9]. Various metal powders
were ball milled for extended time periods in water and in water
saturated with oxygen. The metals that are reduced in size because
of oxide formation at freshly cleaved surfaces have free energies of
oxide formation more negative than the free energy of formation of
water. Silver, copper, and nickel, which do not meet this requirement,
actually show particle growth during the ball milling due to welding.
The addition of molecular oxygen to the ball mill enables copper and
nickel to be reduced since water is no longer necessary for oxide

formation. The free energy of oxide formation in silver is too low
for sufficient reaction to occur to exceed the rate of particle welding.

Particles with diameters below 1 μm are finding increasing use in
metallurgical research [10,11]. They are especially valuable in the
development of dispersion-strengthened materials by powder metallurgy.
They have lower sintering temperatures and more uniform compaction
properties than the same metals in larger particle sizes, but these
possible advantages are easily offset by the difficulty in handling
submicrometer powders. For example, they do not flow in any standard
equipment, they agglomerate easily, the loose powder density is very
low, they oxidize readily in air and are difficult to reduce without
bonding together, and they change their properties when stored.

Even commercial powders only produce good compaction properties
when handled properly. The most common problem is segregation during
transport. Some form of blending or mixing is almost always required
prior to compaction to assure as homogeneous a product as possible.
Blending is required in the preparation of alloys from elemental powders,
in the production of dispersion-strengthened metals, and even in the
preparation of effective sintering mixtures through control of particle
size. Many disappointing results in materials prepared from powders
come from incomplete understanding of the mixing process. Some problems that can occur during mixing are: changes in particle-size distribution through grinding or agglomeration, oxidation of particle
surfaces, segregation of particle sizes during removal from the mixer,
or difficulties in obtaining a representative sample. Fundamental
investigations of mixing require a careful analysis of the frictional
conditions in the powder mass that affect the relative movement of
the particles during mixing.

III. POWDER COMPACTION

The widest industrial use of powder metallurgy is in the compaction of loose powder in a die to form a green compact, which is sintered to yield the final part. Coining, sizing, heat treatments,

TABLE 3

Test Results of Comminuted Metal Powders and Free Energy of Formation
of Metal Oxides and of Water [9]

| Material | Initial particle size, μm | Time, hr | Powders ball milled in pure water | | |
			Oxygen content g oxygen g metal	Surface area, m^2/g	Calculated particle size,[a] μm
Zirconium	6.25	384	0.059	12.1	0.052
Tantalum	0.43	384	0.048	4.42	0.062
Chromium	4.63	384	0.064	9.65	0.058
Iron	1.34	38	0.015	3.55	0.14
430 Stain-less steel	1.36	384	0.14	83.8	0.0065
Nickel	0.90	174	0	0.99	0.45
Copper	3.20	205	0	0.096	4.7
Silver	1.36	237	0	0.16	2.4
Water	----	---	----	-----	-----

[a]Calculated from formula: particle size $(\mu m) = 4/\rho S$, where ρ is
density (g/cm^3) and S is specific surface area (m^2/g) of material.

infiltration, and machining may follow sintering to yield the desired
dimensional characteristics and strength. The press and sinter operations are performed on high-speed presses capable of up to thousands
of parts per hour and in continuous furnaces. The parts may range in
size from less than an ounce to several pounds. Figure 3 shows a
typical high-speed press. Almost one half billion dollars worth of
parts are produced each year on this type of equipment.

| | Powders ball milled in water pressurized with oxygen | | | Oxides or hydroxides | |
Time, hr	Oxygen content, $\dfrac{\text{g oxygen}}{\text{g metal}}$	Surface area, m^2/g	Particle size,[a] μm	Formula	Standard free energy of formation at 298°C, cal/(g-atom oxygen)
-	-	-	-	ZrO_2	-123,900
-	-	-	-	Ta_2O_5 TaO_x	-91,300.x(approx.)
-	-	-	-	Cr_2O_3 CrO_2 CrO_3	-84,400 -65,000 -40,300
-	-	-	-	FeO $FeO \cdot Fe_2O_3$ Fe_2O_3	-58,670 -60,800 -59,100
-	-	-	-	$FeO \cdot$ $(Fe,Cr)_2O_3$	-60,800(approx.[b])
174	0.092	7.72	0.058	NiO $Ni(OH)_2$	-50,600 -54,150
52.5	0.12	36.4	0.012	Cu_2O CuO	-35,450 -30,850
237	0	0.060	6.4	Ag_2O Ag_2O_2	-2,500 -3,300
-----	-----	-----	-----	H_2O	-56,720

[b]This ΔF was assumed to be the same as that of $FeO \cdot Fe_2O_3$.

Before a part can be produced by a press and sinter operation, extensive research must be done on the characteristics of the powdered metal and in the design of the die. For high-speed press operation, the powder must flow at an even and controlled rate into the die. There, it must take on a constant packing density and compress to a green density with a given amount of pressure. The compact must be

Fig. 3. Four hundred ton mechanical compaction press capable
of up to 20 strokes a minute with independent core rod motion which
can be used to activate movable punches or auxiliary die elements.
This press is over 16 ft high and weighs 28 tons. (Courtesy of A C
Compacting Presses, Inc.)

free of voids and irregularities in density, and it must have sufficient green strength to allow it to be handled without chipping or
cracking before it is sintered. During sintering, it must attain a
prescribed final density and retain a known dimensional tolerance for
final processing and usage.

The flow characteristics of a metal powder are determined in a
standard device such as the Hall flowmeter. This is simply a conical
container with an orifice of 0.100 or 0.125 in. diameter. The time
in seconds for a weighed sample of powder to flow through the device
is taken as a measure of its flow characteristics. Flow changes with
humidity, which must be taken into account during plant operations.
Vibration also changes the flow rate as well as the packing density
of the loose powder. The flow rate must be determined for each new
batch of powder and precautions must be taken against agglomeration
and sedimentation during storage.

Once the powdered metal flows into the die cavity, it assumes a
packing density. This must be known and it must be uniform since it
determines the depth of die required to produce any set of dimensions
and density. The tap density is a crude but valuable test for the
maximum degree of packing that a powder will attain with only gravity
to compress it. It can be estimated by pouring a known weight of
powder into a graduated cylinder and vibrating the cylinder until a
minimum volume is attained, which is the tap density for the powder.
It is sensitive to segregation occurring during the test either by
particle size or by density. The ratio between the initial powder
density and the tap density is a measure of the compressibility of the
powder.

The mold in which shape is imparted to the metal powder consists
of the simplest case of a thin metal plate (the die) with a cylindrical
hole through it. Two close-fitting punches enter into this die from
each side by hydraulic or mechanical force; they compress powder contained within the cavity. The part formed in this manner may be ejected
by withdrawing one of the punches and forcing the part out with the

other or fixing the other and moving the die down over it. The steps
of filling the die, compressing the powder in it, and removing the
green part are accomplished in fractions of a second on high-speed
equipment. Complex parts require core rods, multilevel dies, and
multiple upper and lower punches operated in sequence to achieve
densities as nearly uniform as possible.

When the metal powder is compressed, several stages of densification
occur, with some overlap, in the following order [12]:

   1. Slippage of the particles without excessive deformation.

   2. Elastic compression at the particle - particle contact points.

   3. Plastic deformation at these points resulting in contact areas.

   4. Growth of contact areas through further plastic deformation
and particle breakage.

   5. Massive deformation of the particles to a point where the
particulate nature is lost.

   6. Elastic compression of the mass as a whole.

Cold welding occurs between the powder particles as they move
across each other, scraping off the oxide layer and exposing clean
surfaces which adhere together. Plastic deformation of the particles
locks them together mechanically if they are sufficiently irregular
in shape and also work-hardens them. The combination of cold welding
and mechanical interlocking produces the green strength in the compact.
The amount of deformation of each particle determines the density.

In large compacts, deformation and work hardening of the outer
layers of powder may occur before densification of the remaining mass
is complete, resulting in pressure cones at each end of the compact.
Another cause of failures like this is die-wall friction. Such friction results in excessive density variation in both the vertical and
horizontal dimensions of the part. These situations can be alleviated
through the use of lubricants mixed with the powdered metal before
pressing. Paraffin and synthetic waxes; stearic acid; stearates such
as calcium, aluminum, and magnesium; and graphite have been used.
Their disadvantages are that they must be added to the powder and then

removed before sintering, adding two additional steps to the process.
They lower the density of the compacted part, and they reduce the
green strength. For large parts, these property reductions are avoided
by placing the lubricant on the die wall. This procedure is not adaptable to high-speed operations, however.

The parts coming out of the pressing operation must endure several
handling steps before being sintered. Chipping and cracking during
this period cannot be tolerated, so test pieces are subjected to tumbling under standard test conditions to determine abrasion resistance.
The transverse-rupture strength of the part is also measured by applying
a calibrated load to the breaking point.

Green compacts of high (99% of theoretical) density can be made
by isostatic compaction. This is done in flexible molds containing
the powder which are pressurized as high as 100,000 psi in pressure
vessels to produce compacts of uniform density. Such parts do not
have the high degree of dimensional accuracy that is attainable through
die compaction. The technique is also not suitable for automation and
is expensive and time consuming. It is useful for the compaction of
brittle materials and refractory metals such as tungsten, molybdenum,
niobium, and chromium which are further worked by extrusion. These
compactions are done hot and achieve such high degrees of strength
that machining is possible without further sintering.

IV. SINTERING

The conversion of a green compact into a usable metallic object
is accomplished by sintering. The compact is heated to from 0.5 to
0.8 times the absolute melting point of at least one of the major
components for sufficient time to convert the weak cold welds and the
mechanical interlocks between the particles into continuous metallic
phases. Sintering is usually accompanied by shrinkage as the porosity
is reduced. This can be held to about 1% where necessary. The density,
electrical conductivity, and ductility of the compact all increase
during sintering.

The atmosphere in the sintering furnace is vitally important in determining the properties of the final part. All metallic parts are sintered in neutral or reducing atmospheres to avoid the formation of any oxide layer between the particles. Vacuum sintering is useful for removing volatile impurities, and the use of reducing gases often removes any oxide formed during powder manufacture or compaction. Activated sintering involves the use of a halide gas which enters into the material transport mechanism occurring during sintering accelerating the process.

The types of impurities which furnace atmospheres remove are of four types: surface oxides or nitrides, physically adsorbed gases, chemisorbed gases, and mechanically entrapped gases. As much as 80% of the gases originally present in a loose powder mass is mechanically entrapped during cold die compaction. Most of this gas slowly effuses from the compact during the 24 hours following compaction, but it may cause cracking in the green compact prior to sintering.

Sintering in a neutral atmosphere does little more than prevent excessive reaction between the compact and an oversupply of oxygen. Air trapped within the pores of the compact remains to react during sintering. Reducing atmospheres are preferred, which not only react with existing oxide films on the particles, but also react with physically adsorbed oxygen and mechanically entrapped air. Chemisorbed gases are difficult to remove under even the most stringent conditions.

The typical reaction occurring during sintering of iron is:

$$FeO \text{ (solid)} + H_2 \text{ (gas)} \rightleftharpoons Fe \text{ (solid)} + H_2O \text{ (gas)}$$

The equilibrium of the gas-phase reaction may be expressed according to the vapor pressures of the components:

$$\frac{P_{Fe} P_{H_2O}}{P_{FeO} P_{H_2}} = K$$

where K is the equilibrium constant. This is approximately equal to:

$$P_{H_2O} / P_{H_2} \cong K$$

The equilibrium constant for each reaction is related to temperature
by the change in free energy ΔG_T:

$$\Delta G_T = -RT \ln K$$

where R is the gas constant and T is in degrees Kelvin. Since the free
energies of each gas in the sintering furnace are known, it is possible
to calculate the best temperatures and gas compositions to maximize
the reducing reactions.

Hydrogen is the preferred reducing atmosphere for sintering but
is too expensive for commercial operations, so catalytically cracked
(dissociated) ammonia is normally used. This gas contains 25% nitrogen
and can be produced with very little moisture. Since water is a reaction product, its presence in the original gas mixture would lessen
the efficiency of the operation.

An even more common reducing atmosphere than dissociated ammonia
is partially combusted hydrocarbon gas. A mixture containing large
amounts of hydrogen and carbon monoxide results from burning natural
gas in an air-poor chamber. Water vapor also results, however, and
it must be removed before the gas mixture can be used. By varying
the air-fuel ratio, the reducing potential of the gas mixture can be
controlled. Some carburization occurs when iron and other materials
having an affinity for carbon are sintered in these atmospheres because
of the reaction:

$$C \text{ (solid)} + CO_2 \text{ (gas)} \rightleftharpoons 2CO \text{ (gas)}$$

$$\frac{P^2_{CO}}{P_{CO_2}} \cong K$$

This is often beneficial to the final product, but it must be carefully
controlled.

Vacuum sintering is the most positive method for removing mechanically entrapped air during sintering, but the vapor pressure of the
metals at the sintering temperature and the traces of gases remaining
still permit reactions to occur. Vacuum sintering is most successful

when the dissociation pressure of the contaminant oxide is high enough
at the sintering temperature that the removal of oxygen will encourage
the reaction:

$$2M_xO_y \longrightarrow 2xM + yO_2$$

Several mechanisms have been proposed to explain the changes occurring in a metal compact during sintering [13]. These changes are the
growth of necks between the particles, shrinkage of the compact, and
consolidation of the pores. Neck growth has been shown to occur
primarily by the diffusion of vacancies away from the neck through
surface diffusion. The pores between the particles act as sinks for
these vacancies. Volume diffusion of the vacancies to interior grain
boundaries within the particles has also been shown to occur and contribute to the shrinkage of the mass. In either case, the vacancy
migration causes the flow of material towards the neck between the
particles.

A second mechanism responsible for neck growth is evaporation and
condensation. The difference between the radii of the particles and
the necks between them causes tensile stress at the neck and a change
in vapor pressure. This has the effect of establishing a vapor
gradient between the particle surface and the neck so that material
evaporates from the surface and condenses at the neck resulting in
neck growth.

The change in porosity during sintering is dramatic. The green
compact may have between 10 and 70% porosity, all of which is interconnected (open) and characterized by sharp, irregular shapes. The
surface free energy of these pores is very high and, during sintering,
it provides a driving force for material transport which converts
these shapes eventually to spherical pores of minimum surface free
energy. Surface tension at the sharp cracks and crevices in the interconnected pores is very high, so these are the first to fill with
material becoming less rounded.

The size distribution of the pores in a green compact is determined
by the particle-size distribution of the original powder. As sintering

progresses, the smaller pores disappear because they are unstable. The effect of this is to increase the mean pore size. The change in pore-size distribution is also dependent on the sintering atmosphere and the temperature.

Liquid-phase sintering occurs when a compact of several elemental or alloy powders is sintered above the melting point of one of the components. It has several advantages including high sintering rates, the attainment of theoretical density through the elimination of porosity, and the sintering of a mixture containing a very high melting component or of two materials not soluble in each other.

Liquid-phase sintering occurs when the liquid phase wets the solid phase. If it doesn't, the solid phase sinters by the normal solid-state diffusion and the liquid phase is repelled from the compact. The criterion for wetting is that the contact angle θ must be less than 90° in the equilibrium state:

$$\cos \theta = \frac{\gamma SG + \gamma SL}{\gamma LG}$$

where γ is the interfacial energy between the gas (G), solid (S), and liquid (L) phases. When $\theta = 0°$, the solid is completely enveloped by the liquid phase, and a maximum sintering rate is achieved.

The interfacial energy between the liquid and solid phases also determines whether grain growth in the solid phase will occur during sintering. When γSL is low, as is the case where complete wetting occurs, the liquid phase usually penetrates the grain boundaries of the solid phase and prevents any grain growth. The solid-phase particles become completely separated from each other and rearrange to a configuration of highest density. The pore size becomes very small and is completely eliminated if sufficient liquid phase is present. This will happen very readily if the solid phase is soluble to an appreciable degree in the liquid phase.

Infiltration is a common method of densification during sintering. It is a type of liquid-phase sintering where the liquid phase is melted into the compact during sintering. Iron parts, for example, are often

sintered with a slug of copper resting on top. At the sintering temperature, the copper melts and is drawn into the compact by capillary
action. The time required for the copper to replace the voids in the
compact is very short, and the remainder of the sintering time is liquid-
phase sintering. Where the contact angle between the infiltrant and
the host is low, the infiltrant may separate previously sintered particles if sintering is carried out too long. Also, if the host is partially soluble in the infiltrant, its structure may be weakened. These
problems may be avoided by delaying the entrance of the infiltrant.
Sintering and infiltrating in separate steps or placing an intervening
block of porous metal between the infiltrant and the compact are ways
to accomplish this.

The infiltration of iron or iron alloys with magnesium produces
composites with good mechanical properties at relatively low density.
They are easily worked and are corrosion resistant. Iron can also
be infiltrated with aluminum by the use of an infiltration autoclave
[14], which forces the molten aluminum rapidly into the iron skeleton
by gas pressure before the solubility of iron into aluminum can weaken
the structure and plug the pores.

Sintering furnaces can be either batch type or continuous. Figure
4 shows an example of a modern continuous furnace. These are more
expensive than batch furnaces but have greater throughput with more
consistent properties. They are limited, however, to residency times
of less than 2 hr for practical reasons and also cannot be used for
vacuum sintering.

In the continuous furnace, the compacts are placed on a moving
belt which can run through several heating and cooling zones. If a
binder or lubricant has been used, a low-temperature zone is desirable
for burning off the organic material. The hot zone is usually limited
to about 1,150°C because of the structural materials available for the
moving belt mechanism. Higher temperatures require the use of metal
or refractory boats which are pushed through the hot zone. The sintering atmosphere is retained in the hot zone by a burning gas curtain

Fig. 4. Continuous 24-in. mesh belt sintering furnace. (Courtesy of the Drever Co.)

at each end, or in the case of the hydrogen atmosphere, the hot zone is elevated above the entrance and exit so that the buoyancy of the hot gas maintains its purity.

V. SLIP CASTING

Several consolidation methods do not resemble the standard press
and sinter operations but are nevertheless important powder metallurgical processes. These include slip casting, high-energy compaction,
powder rolling, extrusion, forging of preforms, and hot pressing.
With the exception of slip casting, these methods do not require a
separate sintering step to develop good mechanical properties.

Slip casting is an old technique carried over from the ceramic
industry. It is useful in refractory metal fabrication or for specialized shapes made from common metals such as copper or nickel or
even alloys such as stainless steel. A slip is made by suspending
the metal powder in a water solution of alginic acid or polyacrylic
acid and adjusting the pH for minimum viscosity. It is then poured
into a plaster mold which absorbs the liquid. For hollow objects,
the mold can be inverted after filling leaving only a thin film of
the slip on its walls. This is allowed to harden after which it is
removed from the mold and dried in an oven. Sintering is done in a
reducing atmosphere, holding the temperature around 400°F to burn off
the binder before raising it to the sintering temperature of the metal.
Final product densities are in the range of 70 to 80%.

Slip casting is expensive and slow. It is therefore mainly used
for specialized parts of refractory metals or metal-ceramic mixtures
(cermets). NASA, for example, has used slip casting to evaluate fiber-
reinforced composites by forming the matrix metal into a slip and pouring it around the fibers. Also, high-temperature experimental rocket
nozzle parts have been made by slip casting. It is very useful in
experimental programs where the cost of a die cannot be justified.

VI. HIGH ENERGY RATE FORMATION

The utilization of a high rate of compaction in powder metallurgy
produces some very interesting results. The compaction energy can be
applied with a layer of explosive placed around a can containing the

powder or placed above the powder on an open die. Another method
entails the use of gunpowder cartridges of various calibers, which
impart a high velocity to a slug which impacts on the upper piston in
a die containing the powder.

The amount of pressure attained during explosive compaction is
generally of the same magnitude as in conventional presses. Also,
the final density of the explosively compacted metal is no higher
than the conventional product. Yet there are major differences in
the properties of these, which explain the increasing interest in
this research. In 1952, Kennametal, Inc. conducted an experimental
program to compact a titanium alloy using a surplus 14-in. naval gun
barrel partially filled with water [15]. The powder was placed in
waterproof bags and sealed inside the barrel along with an explosive
charge. The detonation created a pressure of 50,000 psi which instantaneously compressed the powder to uniform fully dense ingots.

Presses that are cartridge activated utilize the kinetic energy
of a slug moving at between 100 and 200 m/sec to provide the compaction force. Iron powder compacted in this manner achieves up to 95%
theoretical density with one shot. With a simple die, there is a
noticeable decrease in the amount of force required to eject the compact from the die compared with a conventional press. Experiments
have shown that 2.5 times less pressure is required with compacts of
equal density [16]. One explanation may be the sudden heating and
consequent expansion of the explosively compacted piece. Also, during
explosive compaction, there is very little lateral flow of material.
For the latter reason, rigid dies are not always necessary and in fact
large sheets of metal powder have been compacted in air without a die.

The mechanical properties of compacts formed by high energy rate
compaction are always higher than those of compacts of the same density
made on conventional presses. For example, the electrical conductivity
of the green compact is higher by 20 to 40% and the impact strength
is increased by 12 to 15%. The sintered compacts show proportional
increases in their mechanical properties.

The extreme rate of speed at which the particles are traveling
when pressed together causes local plastic deformation at the contact

surfaces accompanied by a large increase in temperature, which does
not have time to dissipate into the bulk of the particle. The plastic
deformation exposes clean surfaces, and the high temperature at the
contact points encourages particle welding to such a degree that grain
growth across the original particle boundaries can be seen. The higher
the velocity of compaction, the more inter-particle grain growth. Compacts made of spherical copper powder compressed with a detonation
velocity of 7,000 m/sec have mechanical properties approaching the
solid metal. The oxide films between the particles are completely
disturbed, and there is so much plastic flow that the original particles are scarcely recognizable. Conventionally pressed samples compacted at 12,000 psi have very little green strength, and the oxide
films surrounding the particles are largely intact.

High energy rate compaction yields strong compacts with nearly
theoretical densities without sintering. However, there has been very
little commercial interest in this technique so far. As a research
tool, on the other hand, it is very useful. Compacts can be made of
mixtures of metals without regard for melting points or wetting angles.
Titanium with magnesium, or iron with lead are examples of mixtures
not compactable in any other way. Porous filters with precise external
dimensions can be made by placing a powder mixture of the desired metal
and a leachable metal inside a rigid die with an explosive charge in
the center. The resulting compact is then solvent leached to produce
the filter. A potential commercial application is the production of
iron printing characters of high densities and precise tolerances without sintering. Extensive research has shown the feasibility of this
process.

VII. POWDER ROLLING

The practicality of rolling metal powder into strip has long been
intriguing to the powder metallurgist. The conventional procedure
requires that a hot billet be rolled back and forth numerous times to

achieve the desired thinness. With the powder process, one pass through
the rollers followed by sintering produces an endless strip of metal
in a wide range of gauges and compositions. While the efficiency of
this process is obvious and its workability has been proven on the
pilot plant scale, the price difference between raw iron billets and
iron powder is too large for powder rolling to be a viable industrial
practice except for special products.

In powder rolling, the metal powder is fed into the gap between
two rollers where it is compacted into a continuous strip. The friction between the rollers and the powder controls the amount of powder
carried into the gap for compaction. However, for this friction to
carry sufficient powder into the gap, the diameters of the rollers
must be between 50 and 150 times the thickness of the desired strip.
Also, the strip thickness is limited by the angle of friction between
the roller surface and the powder, and the maximum thickness that has
been successfully made has been around 0.25 in. Smaller rollers have
been used for fairly thick strips through the use of "pull strips."
These are two endless belts of metal that run between each roller and
the metal powder. The angle they form at the entrance to the roller
gap is fixed by guides so that it is much less than that formed by
the rollers and gives the effect of much larger rollers. The strips
are pulled through the rollers by the friction of the powder being
compressed between them and the rollers. The surface of the strips
must pick up sufficient powder, however, in order to provide good
compression in the roller gap.

The speed at which powder can be rolled is limited by the gas
being expelled by the loose powder as it is compressed between the
rollers [17]. As the metal powder enters the area between the rollers
(see Fig. 5), it remains uncompacted until it reaches AB which is
determined by the gripping angle α of the rollers. At this point,
compaction begins and continues until the powder reaches the point
of minimum gap CD between the rollers. The powder may be considered
as being in two areas: an incoherent zone above AB and a coherent
zone defined by ABCD. In the coherent zone, the porous powder compact

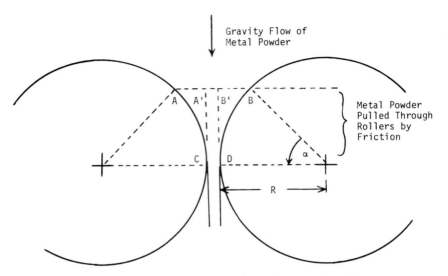

Fig. 5.   Simplified schematic of powder rolling process.

ABCD is being compressed to a dense compact equivalent to A'B'CD.
The volume change, ABCD - A'B'CD, is the volume of gas which is expelled from the powder. This is equal to:

$$V = 2R^2 \sin \alpha \ (1 - \cos \alpha)$$

This volume of gas is being generated at a rate determined by the
roll speed, and it must escape through the coherent/incoherent zone
AB. As it escapes, it fluidizes the incoherent mass of powder. Thus,
the factor limiting the speed of powder rolling is the critical fluidization velocity of the powder. This can be determined experimentally
and is a function of particle size, shape, and mass. At speeds above
the limiting speed, insufficient powder is transported into the coherent zone, and a discontinuous strip is produced.

The speed limitation in powder rolling is a serious drawback to
commercial development. Possible solutions include rolling in atmos-

pheres of lesser densities than air or at lower pressures. Another
possibility is to roll sintered preforms instead of loose powder.
Although the latter will not produce the continuous rolls utilized
in industry, economic factors already rule out any PM process for the
forseeable future. For specialty alloys, however, experiments have
shown that high tensile strength steel sheets can be produced in one
pass using lightly sintered preforms [18]. These preforms, in slab
shapes, can sustain large thickness reductions (up to 70%) by hot
rolling as high as 1,000°C to yield coherent sheets with properties
comparable to conventionally wrought alloys of the same composition.

VIII. EXTRUSION

Some of the high-strength alloys cannot be cast without the formation of segregated brittle phases. When they are consolidated from
alloy powders, uniform billets can be prepared, provided reactions
between oxygen and sensitive components such as chromium can be pre-
vented and if sufficient density can be achieved. The reason for the
latter provision is that the alloys are often prepared by atomization,
which yields low green-strength powders. Hot extrusion is the preferred method of consolidation, which gives full density.

The extrusion process usually starts with an isostatically pressed
cylinder of the alloy powder. This is placed into a mild steel can
and welded shut with a heavy steel plug. The canned alloy powder can
be heat treated to remove impurities by circulating hydrogen through
the compact via vents in the plug. Alternatively, the can can be evacuated through the vents. Prior to extrusion, the can is heated to a
sintering temperature. It is then rapidly forced by a high-pressure
ram through a die which reduces its diameter by a ratio of up to 15
to 1. The resulting rod is essentially fully dense and is about 75%
usable alloy after the canning metal is machined away.

IX.   FORGING OF POWDER METALLURGY PREFORMS

In conventional forging, a desired shape is developed by deforming
metal billets with successive compressive blows of a hammer, several
strokes of a press, or a few passes through a roll forging press.   The
metal billet is usually heated above its recrystallization temperature
before it is formed, which for ferrous alloys is about 1,200°F.   As
deformation of the crystal structure of the billet takes place, new
crystals are nucleated and thus work hardening is avoided.   The resulting piece is fully annealed.   Cold forging is also done and has
the advantages of closer tolerances and better surface finishes because
of the absence of heat and oxidation.   Cold-forged parts, however,
have high degrees of cold-worked stresses in them.

The most common forging process consists of a series of operations
that are done in a single multichambered die.   The billet is first
edged and fullered, which gathers the metal into the areas where the
thickest parts of the finished piece will be and also improves the
grain structure.   Then, there is a blocking step which forms the rough
shape and starts to create flash, followed by a finishing impression.
The flash is trimmed on a second press and a considerable amount of
hand finishing is required.   The entire operation is accomplished by
a skilled operator who adjusts the hammer pressure and moves the billet
from cavity to cavity in the die as the shape progresses.   Any reduction in the number of steps involved increases the rate of die wear
because of the enormous force that is needed to deform a solid metal
billet.

The use of powder metallurgy preforms introduces two major changes
into the process.   First, by pressing and sintering a porous preform
in the general shape of the final part, material waste is almost
eliminated.   Secondly, the porous compact requires a much lower pressure to bring it to the final shape so only a single forging step is
required.   Other advantages of the process are:

1.   The ability to forge parts larger than heretofore possible.

2.   Randomly oriented, fine grain structure.

3.   Equiaxed orientation of physical properties.

4. The ability to begin with chemically pure metals.

5. The elimination of much or all final machining.

6. A good surface finish.

7. Lower forging load.

8. The ability to form complex components in one forging operation.

9. High production rates with semi-automated equipment.

Figure 6 shows a comparison of PM preform forging with conventional forging. By bringing a sintered preform to the forging press, the operator has only one operation to complete per part, increasing his output four times. The preform itself can be made on a semi-automatic press and the number of dies required is reduced since only one is necessary to make the preform and one to forge the part.

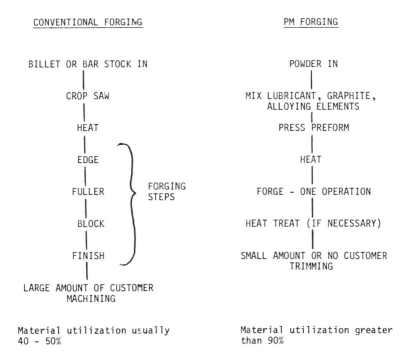

CONVENTIONAL FORGING	PM FORGING
BILLET OR BAR STOCK IN	POWDER IN
CROP SAW	MIX LUBRICANT, GRAPHITE, ALLOYING ELEMENTS
HEAT	PRESS PREFORM
EDGE	HEAT
FULLER } FORGING STEPS	FORGE - ONE OPERATION
BLOCK	HEAT TREAT (IF NECESSARY)
FINISH	SMALL AMOUNT OR NO CUSTOMER TRIMMING
LARGE AMOUNT OF CUSTOMER MACHINING	

Material utilization usually
40 - 50%

Material utilization greater
than 90%

Fig. 6. Comparison of forging sequences between powder metallurgy preform forging and conventional forging (after Ref. 19).

In addition to these economic factors, there is also a quality factor. The metal movement during forging of a sintered preform is considerably reduced, yielding a final part with more uniform grain structure compared to the larger-oriented grain structure common to forged cast metals. The porosity of the preform enables it to be more easily moved in the beginning of the stroke. Only at the bottom of the stroke is full density achieved, which consolidates the last bit of porosity after the final shape has been completed.

A preform slug made from powder must be shaped so that in forging, the material is evenly distributed in the die cavity giving a uniform density throughout the part. Too little powder yields a low density or undersized part; too much powder produces excess flash or high forging pressure. Excessive pressure and die-wall contact lead to short die life. Other factors also influence die life, such as the die material and the lubricant used in forging.

The porosity of the powder metal preform has both advantages and disadvantages in the forging process. One disadvantage is that a gas atmosphere is required to protect the preform during forging. There are two reasons for this: the first is that the metal powder preform, if not protected, would oxidize throughout from the heating step prior to forging, and the second is that scale or decarburization would be objectionable. The lower forging pressures required more than compensate for the inert gas atmosphere. Also, in some cases, forged sintered preforms exhibit higher strength than the equivalent wrought parts. Forged PM gears often show better fatigue life, for example. Some of this is explainable by the isotropic grain structure characteristic of forged sintered preforms.

X. DISPERSION-STRENGTHENED COMPACTS

In most powder metallurgy processes, the existence of an oxide film on the particles prior to sintering impairs the processes of adhesion, surface diffusion, and volume diffusion. To prevent this,

it is usually necessary to work in a reducing atmosphere. The oxides
of some metals such as aluminum and beryllium cannot be reduced in
gas-phase reactions, and since aluminum forms an oxide skin almost
instantaneously upon contact with the air, for many years it was not
considered suitable for powder metallurgy. In 1950, however, a sintered
aluminum material was discovered in Switzerland which had superior
strength properties to cast or wrought aluminum [20,21]. Not only was
the new material an exception to the idea that the oxide layer is always detrimental, but more important, for the first time, a powder
metallurgy product had better properties than did the wrought metal.
The new product was made from ordinary pigment-grade aluminum flake
and contained up to 13% oxide. It had a tensile strength of around
14,000 psi at 500°C and maintained it for over 700 hr. Compared to a
maximum of 3,000 psi at 500°C for conventional aluminum alloys, the
new material was superior to any other, including precipitation alloys
hardened at temperatures above 200°C.

The new material, which contained sintered aluminum, was named
SAP (sintered aluminum powder) and was a member of a new class of
materials. Its production involved a critical milling step during
which the aluminum flakes are further oxidized and agglomerated
to a coarse (100-μm) powder containing an internal dispersion of
oxide (6 to 14% Al_2O_3). This powder is cold-compacted and the
compacts wrapped in aluminum sheet before sintering in air at
500°C. At this point the compacts have no unusual strength properties,
and the bonds between the particles are mostly Al to Al_2O_3. A final
step involving extensive shear is necessary to develop the full
strength of the material. Hot pressing, for example, followed
by extrusion develops a high percentage of Al to Al bonds and deforms
(work hardens) the crystal structure. This, however, only explains
how the sintered aluminum could have strengths comparable to cast
aluminum.

The increase high-temperature properties of SAP are due to the
fine dispersion of Al_2O_3 particles in the aluminum matrix. Most metals
and alloys undergo recrystallization at some characteristic temperature
after which they lose most of their strength properties. This occurs

when the energy of work hardening built up by deformations of the
original crystal structure is released. Above the recrystallization
temperature this energy dissipates leaving a weak strain-free grain
structure. SAP's of more than 7% oxide do not recrystallize until
very near the melting point and consequently retain their cold-worked
strength to that temperature [22].

Recrystallization as well as any form of plastic deformation of a
metal requires the movement of vacancies through the atomic lattice
to positions of less strain. The dispersoid particles in SAP seem
to block this movement, thereby raising the recrystallization temperature and increasing the yield strength. The number of particles and
their spacing in the matrix should affect these properties more than
the volume percent of dispersoid, which has been proven experimentally
[23].

The relationship between the volume fraction of the dispersed
phase, the particle size of the dispersed phase, and the interparticle spacing is given by [24]

$$IP_{av} = 4\left(\frac{V}{S}\right)_{dp} \left(\frac{1}{VF_{dp}} - 1\right)$$

where IP_{av} is the average interparticle spacing, VF, is the volume
fraction of the dispersed phase, V the total volume of the dispersed
phase, and S the total dispersed phase surface. $\left(\frac{V}{S}\right)_{dp}$ is a measure
of particle size and is related to particle diameter for spherical
particles by:

$$\left(\frac{V}{S}\right) = \frac{diameter}{6}$$

This relationship provides a guide to the maximum dispersed-phase
particle diameter that can be used to produce a desired interparticle
spacing. For example, if an interparticle spacing of 1 μm is sought,
15% of a 0.26-μm dispersoid might be used. Higher amounts of a coarser
dispersoid could be used, but percentages exceeding 15% generally reduce
strength properties.

SAP is a nearly ideal example of a dispersion-strengthened material. The formation of oxide cn aluminum powder in extremely thin (100-A) layers results in a very fine dispersion with the interparticle spacing about 0.4 μm. Agglomeration of the oxide particles does not occur, and the oxide itself is one of the most chemically and thermally inert materials known. Few other structurally desirable metals have usable oxides, for reasons of reactivity or solubility at elevated temperatures. Therefore, strengthening other metals requires different dispersoids and dispersion methods.

Tungsten lamp filaments have been made by powder metallurgy for half a century and have been strengthened by a dispersion of thorium to achieve a finer grain structure and better ductility. The capacity for dispersion strengthening was considered a unique property of tungsten, and it was not until the discovery of SAP that a generalized interest in dispersion strengthening started. Since that time, dispersion-strengthened lead, nickel, and nickel-chromium have become commercial products.

Table 5 shows the range of metals which have been dispersion-strengthened by powder metallurgy techniques. The high-temperature strengths of these materials depend on the stability of the dispersoid. It should not react with the matrix, and it must have a higher melting point. Oxides such as thoria (ThO_2), alumina (Al_2O_3), beryllium oxide (BeO), and magnesium oxide (MgO) have these properties and have been used in most dispersion-strengthened composites. The carbides used in precipitation-hardened steels do not have good high-temperature properties. They are not stable and migrate or dissolve at elevated temperatures.

The production of all dispersion-strengthened metals involves the following:

1. The use of PM techniques for the preliminary preparation of the alloy.

2. A hard insoluble particulate as the dispersed phase.

3. Extrusion or other high energy, hot working procedure for consolidation.

TABLE 4

Dispersion-Strengthened Materials Developed by Industry
and Research Institutions (from reference 25)

Matrix	Dispersoid
Aluminum	Alumina,[a] iron aluminide
Beryllium-copper-tin	Beryllium oxide
Chromium	Magnesium oxide
Cobalt	Thoria, various oxides
Cobalt-chromium	Various oxides
Copper	Silicon dioxide, alumina, beryllium oxide[a]
Inconel	Various oxides
Iron	Beryllium oxide
Iron-chromium	Thoria
Iron-nickel-chromium	Thoria
Lead	Lead oxide[a]
Nickel	Alumina, thoria[a], various oxides
Nickel-chromium	Alumina, thoria[a]
Nickel-cobalt	Various oxides
Nickel-molybdenum	Thoria
Niobium	Various dispersoids
Silver	Silver oxide
Stainless steel	Various carbides, thoria, beryllium oxide
Tantalum	Various dispersoids
Tungsten	Thoria
Vanadium	Various oxides
Zinc	Various oxides
Zinc-copper	Titanium
Zirconium	Beryllium, various oxides

[a]Production item.

Various methods have been used to dispersion-strengthen metals
which do not have suitable oxide layers. The most common method
involves the mechanical blending of a suitable dispersoid powder with
the matrix powder. The matrix must be less than 10 μm in diameter,

preferably less than 1 μm. The dispersoid should be in the range of 1/30 to 1/250 the diameter of the matrix particles in order to achieve the best distribution [26]. Special ball-milling techniques have produced oxides as fine as 0.01 μm for this purpose [27]. From 0.5 to 15% by volume of the dispersoid is blended into the matrix powder before compaction. This blending operation is critical to the strength of the final composite since agglomeration of the dispersoid and segregation due to the differences in particle size and specific gravity of the two phases are likely to occur.

Other methods for producing powders for dispersion-strengthened composites include:

1. Precipitation from the melt of an intermetallic phase. The atomization of fused aluminum containing dissolved iron is an example. A powder containing finely dispersed $FeAl_3$ results [28].

2. Internal oxidation. Powders of alloys made up of a small percentage of an easily oxidized metal in a more inert metal can be treated with heat and oxygen to precipitate oxide crystals within the particles. Examples are aluminum in copper, chromium, or nickel [29].

3. Preferential reduction of oxides. Powder mixtures of oxides such as NiO or MoO_2 with Al_2O_3 can be easily comminuted to a fine particle size and then reduced prior to compaction. This produces a very fine dispersion of alumina in a nickel or molybdenum matrix.

Dispersion-strengthened composites are an example of the ability of powder metallurgy to "custom make" metallic materials, attaining superior properties that cannot be achieved through fusion metallurgy. Dispersion-strengthened composites are used in applications requiring high-temperature strength, which would otherwise require more expensive refractory metals. Examples are aircraft hydraulic tubing and turbine engine rotor buckets. Dispersion-strengthened lead enables radiation shielding to be formed into structural shapes having less mass than ordinary lead. The use of dispersion-strengthened composites is limited at this time partly because the composites cannot be welded without agglomerating the dispersion and weakening the joint. Nevertheless, their unique strength properties assure them an important place in the future of metallurgy as more sophisticated applications call for maximizing strength while minimizing expense and mass of materials.

XI.   SAFETY IN HANDLING POWDERS

The particle characteristics of some common grades of iron powders are shown in Table 6, which compares the size distribution of each powder with its average particle size. Unfortunately, however, the standard mesh sizes are only good for particle sizes down to 44 μm (325 mesh). Those particles retained by this screen are not inherently dangerous because of particle size. They settle quickly from the air and are easily retained in processing equipment. A 50-μm particle with a specific gravity of 10 has a terminal settling velocity of over 10 fps. This is lowered to 1 fps for a 10-μm particle and to 0.01 fps for a 1-μm particle. Thus, the finest particles in a powder may remain airborne for a considerable length of time if dispersed. These fine particles are dangerous from several aspects.

The dangers inherent in handling fine powders can be classified in the following groups: (1) respirable toxicity, (2) contact toxicity, (3) flamability, and (4) explosivity. These all result from the particle size of each material and are not properties of the materials themselves. A 1-lb piece of beryllium, for example, will not explode or burn. It is not poisonous, and it can be safely handled without gloves. Reduced to a fine powder, however, beryllium may ignite spontaneously or even explode if dispersed in the air. It is highly toxic if inhaled and no more than 0.002 mg/cu m is permitted in workroom atmospheres by the ACGIH (American Conference of Governmental and Industrial Hygienists).

The -325-mesh column in Table 6 takes on more meaning when it is realized that this is the lower limit of a particle-size distribution. The 78-μm electrolytic iron powder poses little problem, having only trace amounts of -325 mesh particles. The 6-μm reduced iron and the 7-μm carbonyl iron powders, on the other hand, are almost exclusively in this size range. They can be expected to liberate substantial

TABLE 5

Characteristics of Three Types of Iron Powders [30]

Type of powder	Average particle size (μm)	Particle size distribution (mesh)				Specific surface area (m³/g)	Apparent density (g/cm³)
		+150	-150 +200	-200 +325	-325		
Electro-lytic	78	29.26	16.39	54.16	-	265	3.32
	63	21.98	16.00	50.10	10.0	452	2.56
	53	2.38	21.0	74.0	3.0	1,150	2.05
Reduced	68	28.5	15.5	54.5	1.0	516	3.03
	51	-	6.5	81.5	12.0	945	2.19
	6	3.5	2.0	13.5	78.5	5,160	0.97
Carbonyl	7	2.5	0.1	1.0	95.5	3,460	3.40

quantities of airborne particles in any mixing, milling, or transfer
operation; and appropriate shielding and venting procedures must be
used. Also, any form of particle movement causes some comminution,
reducing the average particle size and liberating more fines.

The respirable range of particle sizes is approximately 1 to 5
μm. These particles are not trapped in the nose or throat and therefore reach the lungs where they can be absorbed directly into the
bloodstream. Finer particles tend to remain suspended in the air-
stream and may be exhaled. Metals such as aluminum, beryllium, iron,
lead, nickel, and zinc are dangerous if inhaled as fine powders. Iron
dust, for example, can cause conjunctivitis, choroiditis, retinitis,
and siderosis of tissues if iron remains in these tissues [31]. The
constituent elements of alloy powders must also be considered when
reviewing possible hazards. Brasses can contain lead, antimony, and
even arsenic. The current maximum safe concentrations of some common
materials are given in Table 7.

Exposure to the skin of fine metal powders must also be considered.
In cases where significant dermal reactions from specific materials
have been reported or suspected (e.g., with nickel), care should be
taken to minimize skin exposure, especially by those individuals who
have previously shown dermal hypersensitivity. In many cases, a
manufacturer must simply be alert to employee complaints since little
literature in this area is available. (One of the most recent [33]
has yet to be translated into English.)

Many fine metal powders are inherently flammable and should be
protected from open flames, sparks, or chemical oxidizing agents.
Some metal powders such as chromium may ignite spontaneously in air
and should be stored under inert gas. Handling equipment should be
electrically grounded to avoid sparks resulting from static charge
buildup on powders. Supplies of powdered graphite, dolomite, or
sodium chloride-not water-should be available to quench fires
wherever fine powders are handled.

The explosive hazard of finely dispersed powders is generally well
known (for example, reference 34). Extreme care must be taken during

TABLE 6   Threshold Levels of Contamination

Metal	Current ACGIH limit[a] (mg/m^3)	Special caution[b]	Suggested TLV for Ultra fine Dusts[b,c] (mg/m^3)
Aluminum			2.0
Beryllium	0.002		
Chromium[d]	0.1 (chromic acid and chromates - as CrO_3)	Carcinogenic potential sensitization	0.002
Cobalt[d]	0.1	Sensitization	0.002
Copper	0.1 (fume) 1 (dusts)		
Ferrovanadium	1 (dusts)		
Iron oxide	10 (fume)		
Iron pentacarbonyl	0.08		
Iron	1 (salts - as Fe)		
Lead			0.15[a]
Magnesium oxide	10		2.0
Molybdenum	5 (soluble) 10 (insoluble)		0.5
Niobium			1.0
Nickel carbonyl[d]	0.007	Carcinogenic potential sensitization	0.002
Nickel[d]	1		
Silver	0.01		
Thorium oxide		Carcinogenic potential (radiotoxicity)	0.002
Tin	2		
Tungsten	1 (soluble) 5 (insoluble)		0.5
Uranium	0.2 (salts - as U)		
Zirconium[d]	5 (salts as Zr)	Sensitization potential	0.1

[a]From Threshold Limit Values for Chemical Substances and Physical Agents in the Workroom Environment with Intended Changes for 1972, American Conference of Governmental Industrial Hygienists, Cincinnati, 1972. [b]From Ref. 32. [c]Concentration of metal element. [d]Avoid skin contact.

transfer of powders with appreciable fine fractions. As much as one pound of iron powder may remain airborne after pouring 100 pounds of a powder mixture containing only 1% of particles less than 5 μm in diameter. This poses an explosive hazard. Good ventilation, or inert gas blanketing is necessary in any powder transfer operation.

REFERENCES

1. A. B. Backensto, in Perspectives in Powder Metallurgy, Vol. 3, H. H. Hausner, K. H. Roll, and P. K. Johnson, Eds., Plenum Press, New York, 1968.

2. G. I. Friedman, Int. J. of Powder Metallurgy, 6, 43-55 (1970).

3. P. V. Gummeson, Powder Metallurgy, 15, No. 29, 67-94 (1972).

4. W. D. Jones, Fundamental Principles of Powder Metallurgy, Edward Arnold, London, 1960, p. 224.

5. F. Lochmann, in Metallische Spezialwerkstoffe, Vol. 7, Series A, Berlin, 1963.

6. W. L. Batten and G. A. Roberts, (to Vanadium-Alloys Steel Co.), U.S. Pat. 2,956,304 (1960).

7. B. C. Bradshaw, J. Chem. Phys., 19, 1057-1059 (1951).

8. A. Arias, NASA TN D-4912 (1968).

9. A. Arias, NASA TN D-4862 (1968).

10. W. E. Kuhn, Ed., Ultrafine Particles, Wiley, New York, 1963.

11. C. G. Goetzel and M. A. Steinbery, in Modern Developments in Powder Metallurgy, Vol. 1, H. H. Hausner, Ed., Plenum Press, New York, 1966.

12. P. J. James, Powder Metallurgy International, 4, 193 (1972).

13. D. L. Johnson, in Modern Developments in Powder Metallurgy, Vol. 4, H. H. Hausner, Ed., Plenum Press, New York, 1971.

14. R. Kieffer, G. Jangg, and F. Csősz, Powder Metallurgy Int., 3, 182 (1971).

15. Anonymous, The Iron Age, 170, No. 12, 56 (1952).

16. O. V. Roman, in Modern Developments in Powder Metallurgy (H. H. Hausner, ed.), Vol. 4, Plenum Press, New York, 1971.

17. P. E. Evans, in Modern Developments in Powder Metallurgy, Vol. 4, H. H. Hausner, Ed., Plenum Press, New York, 1971.

18. C. H. Weaver, R. G. Butters, and J. A. Lund, Int. J. of Powder Metallurgy, 8, 3 (1972).

19. P. K. Jones, Powder Metallurgy, 13, No. 26, 114 (1970).

20. R. Irmann, Tech. Rundschau, 19, 1-4 (1949).

21. A. V. Zeerleder, Zeitschrift für Metallkunde, 41, 228-231 (1970).

22. G. S. Ansell, Trans. AIME, 215, 249-250 (1959).

23. W. S. Cremens and N. J. Grant, Proc. ASTM, 58, 714 (1958).

24. C. S. Smith and L. Guttman, J. of Metals, 5, 8 (1953).

25. H. W. Blakeslee, Powder Metallurgy in Aerospace Research, National Aeronautics and Space Administration, Wash. D. C., 1971.

26. K. M. Zwilsky and N. J. Grant, Trans. Met. Soc. AIME, 221, 371 (1961).

27. M. Quatinetz, R. J. Schafer, and C. Smeal, Trans. AIME, 221, 1105-1110 (1961).

28. R. J. Towner, Metal Progress, 73, 70 (1958).

29. L. J. Bonis and N. J. Grant, Trans. Met. Soc. AIME, 224, 308 (1962).

30. H. H. Hausner, in Handbook of Metal Powders, A. R. Poster, Ed., Reinhold, New York, 1966.

31. N. I. Sax, Dangerous Properties of Industrial Materials, 3rd ed., Van Nostrand Reinhold, New York, 1968.

32. F. J. Viles, Jr., R. I. Chamberlin, and G. W. Boyler, Jr., NASA CR-72492 (1966).

33. J. T. Brakhnova, Toxicity of Powdered Metals and Metal Compounds, Academy of Sciences of the Ukranian SSR, Kiev, 1971.

34. A. G. Alekseev and V. A. Kritsov, Explosive Hazards of Metal Powders, Naukova Dumka, Kiev, 1971.

Chapter 6

PHARMACEUTICALS

J. T. Carstensen

School of Pharmacy
University of Wisconsin
Madison, Wisconsin

I. INTRODUCTION

The solid-gas interface enters pharmaceutical considerations in cases of solid dosage forms, viz., tablets and capsules. It furthermore plays a role in gas sterilization of parenteral powders. These three subjects will be dealt with separately below. Since the reader may not be familiar with pharmaceutical operations, a brief description of and rationale for each process is presented here.

Tablets are produced as shown schematically in Fig. 1. A granulation flows from a hopper at A into a die cavity formed by a die and a lower punch. Subsequent to filling, the tablet is formed (compressed) by lowering an upper punch (point B), and finally ejected by bringing the lower punch up. The movement of the punches is controlled by cams above and below the die table. The tablet is scraped off by the scraper bar D. Dual and triple hoppering is, of course, possible and is used.

243

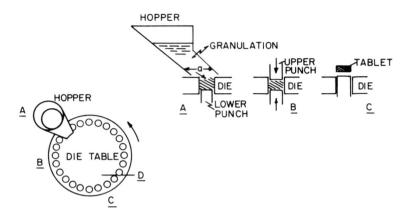

Fig. 1. Principle of tablet formation on a rotary tablet press.
A: Hopper, B: compression point, C: ejection point, and
D: scraper bar. Loose granulation is depicted by short horizontal
lines, compressed matter by slanted lines.

If the length of the hopper opening is a cm, the diameter of the
die is d cm, the depth of the die-punch cavity is h cm, and if the
die table has a diameter of D cm and rotates at Ω rpm then the flow
rate of a granulation of apparent density ρ g/cm^3 must obviously be

$$W = \frac{h\pi^2 d^2 D\Omega\rho}{4a} \quad \frac{g}{min} \tag{1}$$

It is seen, therefore, especially in cases of high-speed machines
with high values of Ω, that flow rates are important.

Flow rates of monodisperse powders are a function of particle
diameter [1]. This is illustrated in Fig. 2, and it is seen that
there is an optimum diameter for flow. For polydisperse powders the
amount of "fines" is of importance [2], as shown in Fig. 3, and for
bidisperse mixtures there is an optimum amount of fines. Polydisperse mixtures show similar optimal compositions [3].

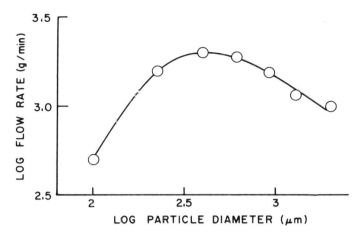

Fig. 2.   Flow rates of monodisperse powders as a function of
particle diameter.   After R. Carr, Chem. Eng., 67, No. 4, 121 (1960).

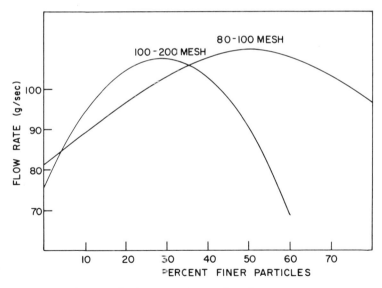

Fig. 3.   Effect of fines on flow of 10/20 mesh lactose granulation
After Gold et al. [2].   Reproduced, with permission, from J. Pharm. Sci.

To increase or optimize flow rates of a powder mixture, parti-cle-size enlargement is often necessary for the reasons cited, and this is conventionally accomplished by granulation techniques. Gran-ulation is accomplished by adding to a powder mixture a solution or suspension of the granulating agent. The latter can be an organic water-insoluble film former such as polyvinylpyrrolidone (in which case isopropanol, e.g., is used as solvent) or it may be a water-sol-uble compound (e.g., gelatin or starch, in which case the solvent is water).

Conventionally the powder to be granulated is placed in a suit-able blender (sigma-type, Pony-type, or V-blender to mention a few) and the granulating liquid is added, e.g., by pumping or direct pouring. The powder "lumps up" and is then dried either by tray drying, fluid-bed drying, or vacuum drying. This drying process is the particular gas-solid interface process involved herein.

II. DRYING

Tray drying leads to conventional drying curves [4]: (a) a con-stant rate period, (b) a falling rate period, and (c) a final drying period. In pharmaceutical granulations, however, the constant rate period is virtually absent, because only a limited amount of solvent is added to the powders. Furthermore, drying into the final drying period is not desirable because it leads to changes in crystal form or particle sizes of hydrates (and usually one or more ingredients is a hydrate) [5] which may affect bioavailability, on one hand, and adversely affect compressibility on the other hand. The drying is, therefore, primarily dictated by the falling rate period, where the relation

$$\frac{dW}{dt} = -K(W - W_\infty) \tag{2}$$

holds. W is the weight of water, ∞ denotes content at end of the falling rate period, t denotes time, and K is a constant given by

$$K = \frac{hA(T_a - T_s)}{\Delta H[W_0 - W_\infty]}$$ (3)

where ΔH is the heat of vaporization of the bound solvent, A is surface area, T is temperature, h is a heat transfer coefficient, and
subscripts a and s denote air and surface, respectively. Inserting
Eq. (3) into Eq. (2), integrating, and imposing initial conditions
yields

$$\log \frac{W_0 - W_\infty}{W - W_\infty} = \frac{Ah\Delta T}{H(W_0 - W_\infty)} t$$ (4)

Knowledge of A, ΔH, W_0, W_∞, T_a, and T_s allows calculation of h. In
the case of aqueous granulation, T_s may be equated to the wet bulb
temperature, so that $\Delta T = T_a - T_s$ is known, and ΔH can be estimated
as the heat of vaporization of water at T_a. Shepherd et al. [6]
have shown that h depends on mass velocity M of air (in lb/ft^2-hr):

$$h = 0.013M^{0.8}$$ (5)

Oliver and Newitt [7] and Pearse et al. [8] believe that the
liquid movement is governed by frictional forces in the sense that
the Kozeny equation is obeyed. In this manner of visualizing the
drying process there is no fundamental difference between the constant
and the falling rate period; the change from the first to the second
simply implies that liquid is no longer transported to the surface
via capillary forces, but that it must be transported by vaporization
in situ and then by vapor diffusion.

Ridgway and Callow [9] checked this concept for consistency by
equating the product of concentration gradient and diffusivity with
the vapor diffusion drying rate. If q is the height of the film located above the bed (which is a cm thick and has a cross section of
A cm^2), P is the saturation vapor pressure, D the vapor diffusivity,
and μ the permeability (so that the effective diffusivity is μD in
the dry part of the bed), then, since in the falling rate period the

concentration gradient operates over the part of the bed that is dry
and the boundary film,

$$\frac{dW}{dt} = \frac{P}{(q/D) + (x/\mu D)} = \frac{DP\mu}{q\mu + x} \tag{6}$$

where x is the thickness of the part of the bed that is dry. At the
base of the powder bed (i.e., when the falling rate period is about
over), the rate will be

$$\frac{dW^*}{dt} = \frac{\mu PD}{\mu q + a} \tag{7}$$

For solvent granulations (as opposed to aqueous granulations) there
is a constant rate period, so that

$$\left(\frac{dW}{dt}\right)_c = \frac{DP}{q} \tag{8}$$

The three drying rates can be obtained graphically, and since P, q,
and a are known, it is possible to calculate D and μ and cross-check
(since there are three equations for two unknowns). There is good
correlation in checks of this sort, and, furthermore, the permeabilities fit well with the theoretical Kozeny permeabilities given by
$\mu = \varepsilon^3/(1 - \varepsilon)^2$, ε being the void fraction. The adherence of Eq. (6)
to experimental data is shown in Fig. 4.

The above considerations make the assumption that drying is uniform from particle to particle. However there is, by necessity, in
granulations a range of granule sizes, and in the strictest sense it
is not permissible to assume that the moisture is evenly distributed
during and after drying. If a granule cut has a particle radius r,
then [10] if $\kappa = r^2/(\pi^2 D)$:

$$\frac{W - W_\infty}{W_o - W_\infty} = \frac{6}{\pi^2} \sum_{j=1}^{\infty} \frac{1}{j^2} \exp\left(\frac{-j^2 t}{\kappa}\right) \sim \frac{6}{\pi^2} \exp\frac{-t}{\kappa} \tag{9}$$

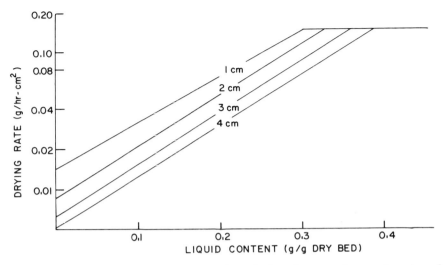

Fig. 4. Evaporation of water from beds of magnesium carbonate of various thicknesses. The thicknesses of the beds are 1 cm, 2 cm, 3 cm, and 4 cm. After Ridgway and Callow [9].

W, hence, will depend exponentially on r^{-2} and the smaller granules will (as is intuitively obvious) be dryer than the larger ones. This is demonstrated in Fig. 5. In the case cited a granulation was dried for 1 hr. The air inlet temperature was 300°F, the air outlet temperature 113°F, and the dryer was operating at 3.5 rpm.

The dried granulation was sieved, and the various sieve cuts subjected to toluene moisture analysis. The percent moisture was then plotted as a function of particle size as shown in Fig. 5A. Separate experimentation showed that $W_\infty = 0.42$, and the plotting of $\log [W - 0.42)/6.58]$ as a function of γ^{-2} produced a straight line (Fig. 5B) with slope $-t'4\pi^2 D$ cm^2; i.e., the diffusion coefficient for the particular situation would be $-(slope)/(4\pi^2 t')$ cm^2/sec and intercept $\ln(6/\pi^2) = -0.5$; t' is here 3,600 sec. It is seen from the graph, that (a) the data adhere fairly well to linearity and that (b) the intercept is -0.56, in good agreement with the theoretical. These two facts substantiate the view that granules dry by a dif-

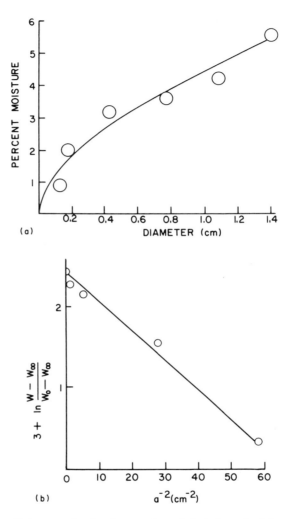

Fig. 5.   Water content of granules dried in a batch as a function of particle size.

fusion process, and that, therefore, moisture content is a function of mesh cut. In cases where granules are not case hardened, the moisture levels equilibrate and the granulation is uniform when it reaches the tablet press; most wet granulations, however, exhibit

case hardening to a greater or lesser degree, and the above phenomena
must be accounted for in processing.

The diffusion coefficient for curve A in Fig. 2 is seen to be
$(3.69 \times 10^{-2})/(4\pi^2 \times 3,600) = 3 \times 10^{-7}$ cm^2/sec, which is of the right
order of magnitude for diffusion through capillary void space [4,9,11].

If, as is most often the case, granules are stored for a while
prior to compression, then equilibration will take place and the granulation will be uniform with respect to moisture at the time it is
fed to the tablet press. In continuous or automatic batch processes,
however, this is not the case. For instance, in the Lewis-Howe automated batch process lack of equilibration has necessitated a screening
and recycling program like that shown in Fig. 6 [11]. Granulation is
made in the mixer by metering, in correct amounts, of solids and granulating liquids. The mixture is then fed into a Bartlett-Snow rotary
dryer 27 ft long and 5 ft in diameter; movement through the countercurrent air stream is accomplished by means of a bucket principle.
The outgoing granulation is sized by a Sweco screen set: the fines
(smaller than 70 mesh) are trapped in a collecting system and returned
to the granulating step; the over-coarse material (over 12 mesh) is

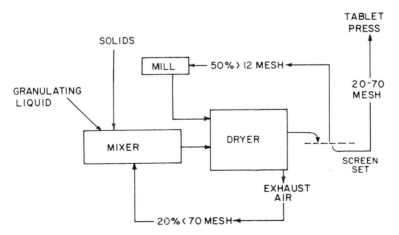

Fig. 6. Lewis-Howe automated batch method. Courtesy of Lewis-Howe
Company, St. Louis, Mo. [11].

fed into a mill and returned to the drying step at A. The 12-70 cut
is lubricated and compressed. The fact that the 12-mesh cut is high
in moisture necessitates the redrying, otherwise it would be simpler
to mill, lubricate, and compress without further reprocessing. The
fines, of course, must be regranulated for particle-size enlargement,
and also because they are too dry.

It follows from Eq. (9) that the finer a granulation, the faster
it will dry, but the optimum cut for compression is usually 10-100
mesh so that particle size is not an adjustable parameter. It should
be noted that the toluene moisture test [12] is the most frequently
used direct means of moisture testing. When it is used for granulations, the dried granulate should be macerated with some of the
toluene in a mortar; otherwise reproducible results are difficult
to obtain.*

There are other automated batch processes, e.g., an automated
vacuum process. Here a V-blender is equipped with vacuum-tight
seals. The intensifier bar is hollow and accommodates (a) access to
vacuum and (b) an inlet for liquid granulating solutions. Models of
about 100 cu-ft capacity are in industrial use; the Patterson-Kelley
[13,14] solids processer is an example of this principle. In this
process the fall of the mass (wet and later dry) through the length
of the mixer arms on each revolution creates new surface exposure
which facilitates the vacuum drying; the mechanical impact, however,
is substantial. Moisture distribution in granulates does not follow
Eq. (9) because of particle-size reduction during drying, i.e., r
is a function of t.

Drying time depends on the air velocity over the wet surface
and on thicknesses of stagnant layers on the surfaces during drying.
In drying of beds, e.g., tray drying, the air passes over the tray
and this, obviously, is not effective in dissipation of stagnant mois-

---

*Other methods used are: loss on drying (e.g. by Cenco moisture
balance or by vacuum at 110°C) and Karl Fischer method; the latter is
often not applicable due to interference of excipients or active component.

ture layers and the preceding section actually dealt with expressions
based on the recession of such fronts. If a powder mass is fluidized,
intimate contact is established between the drying air and the solid,
and this kind of drying is gaining popularity in the pharmaceutical
industry. The design shown in Fig. 7 is characteristic of the approach used: the wet granulation is placed in the portable basket,
which is conical upward; the basket is equipped with wheels to facil-
itate transport. It can be wheeled into place over the air inlet;
a bag collector made of air-permeable material is fitted over it and
air flow started. The dryer then is made to operate by use of sufficiently high air velocity to allow fluidization of the granulation.
Fewer man-hours are needed for this process than for conventional
tray drying. Precautions against dust or solvent explosions have
been made in most of the industrially available designs (e.g., via
proper grounding).*

The pressure drop (ΔP) over the fluid bed is a function of the
weight of the load [15]. Figure 8 shows the relationship between
the logarithm of the pressure drop and the logarithm of the air velocity. At low velocity no fluidization takes place (OQ). With higher
velocity, ΔP increases until it balances gravity (point Q); further
increase in air velocity causes particles to separate (bed expansion),
i.e., at point R fluidization starts. At point S, the bed expansion
equals the depth of the dryer bed, and beyond this point pneumatic
transport takes place.

It is advantageous to monitor the exit air temperature, since
a sudden increase in temperatures indicates completion of drying.
It is possible to automate the monitoring and granulations can be
dried in this fashion to the equilibrium moisture content. Tray
drying procedures frequently lead to overdrying, which is irreversibly
detrimental to the compression qualities of the granulate. The sudden increase in temperature denotes the end of the constant drying

------

*For example, by Fitzpatrick Co., Elmhurst, Ill., and by Glatt
and Aeromatic Driers, Basel, Switzerland.

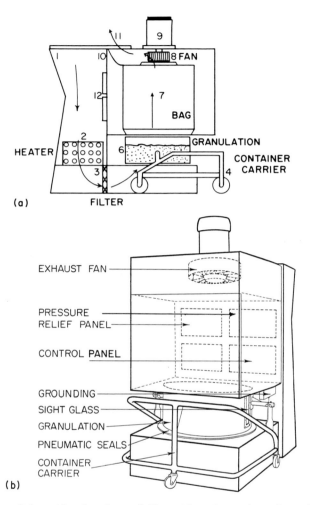

Fig. 7.   Schematic drawing of Fitz-Aire dryer based on fluid-bed
drying.   Reproduced with permission of the Fitzpatrick Co.,
Elmhurst, Ill.

period as shown by Scott et al. [16] and this point corresponds to
the equilibrium moisture content; there is fundamentally no difference
between heat and mass balances in tray vis-à-vis fluidized drying [16].
ΔH in Eq. (3) is the heat required for the total granulation, so, de-

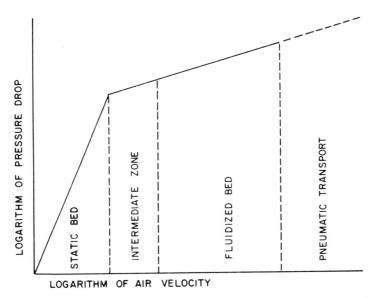

Fig. 8. Pressure drcp as a function of the logarithm of air velocity. After Scott et al. [16]. Reproduced, with permission, from J. Pharm. Sci.

noting by L the latent heat per unit weight, and denoting by m the mass of the granulation, Eqs. (2) and (3) may be combined in the form:

$$\frac{dW}{dt} = \frac{h\Delta T}{\gamma\Delta H} \quad \frac{h\Delta T}{\gamma Lm} \tag{10}$$

Scott et al. [16] tested this by varying ΔT, and found the rate of drying to be a straight line through the origin as shown in Fig. 9a. The logarithmic form of Eq. (10) predicts that

$$\log(dW/dt) = \alpha - \log m \tag{11}$$

where $\alpha = \log(h\Delta T/\gamma L)$; this is observed well as shown in Fig. 9b.

Thermal efficiency can be expressed in the form 100 E_t/E where E_t is the minimum theoretical energy input required for drying (btu/min) and E is the actual energy used, i.e., E = LΔW, where ΔW is the

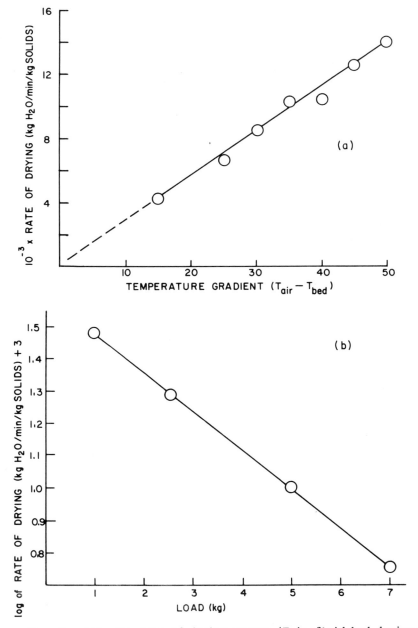

Fig. 9. (a) The rate of drying versus ΔT in fluid-bed drying;
(b) the rate of drying versus the bed load m. After Scott et al.
[16]. Reproduced, with permission, from J. Pharm. Sci.

total weight of water which has been removed from initial time until
attainment of the equilibrium moisture content. Scott et al. [16]
showed that for fluid-bed dryers the efficiency is 2-6 times greater
than for tray dryers.

By placing a feeding nozzle in the fluid bed it is possible to
pump in granulating liquid so that fluid-bed dryers can be modified
to serve as granulating units. This can be extended to a completely
automated granulation unit such as described by Rankell et al. [17]
and the general principle is outlined in Fig. 10. Air is blown via
a damper through a heat exchanger into the fluid bed; powder is metered in (at adjustable rate) below the surface of the fluidized powder and granulation is fed in (via a metering pump with adjustable
rate) through the nozzle at point F. The position of the nozzle
can be altered and the spray is directed downward. The air passes

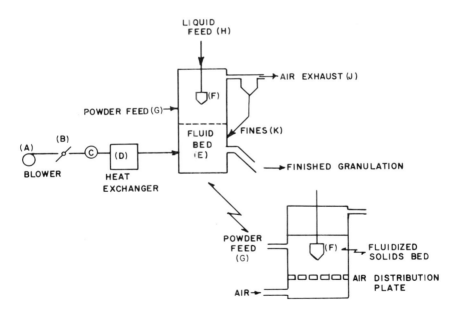

Fig. 10. Continuous granulation set up. Blower (A), damper (B),
orifice controlling and measuring device (C), heat exchanger (D),
fluid bed (E), spray nozzle (F), powder feed inlet (G), liquid feed
(H), air exhaust (J), fines return (K). After Rankell et al. [17].
Reproduced, with permission, from J. Pharm. Sci.

through a cyclone and is exited at J, the cyclone returning fines to the fluid bed. The finished granulation is removed through an adjustable exit valve, and the flow rate out can be controlled by its vertical position. In this manner fluid input, solid input, air flow, and finished output can be adjusted; some start-up adjustments are necessary until steady state is achieved.

The material balance [16] is made, using the following nomenclature:

ω = flow rate of air (on dry basis)
C = solids concentration in granulating liquid
Ω = flow rate of granulating fluid
R = absolute humidity (kg water/kg dry air)
c = moisture content of solids (kg/kg dry)
σ_o = emptying rate of finished product (kg dry/min)
σ_i = feed rate of solids (kg dry/min)
subscripts i and o = inlet and outlet
x = concentration of drug (kg/kg dry, assayed)

The following balances are then required:

$$\sigma_i = \sigma_o \quad \text{(for air)} \tag{12}$$

$$\sigma_i + C\Omega = \sigma_o \quad \text{(for solids)} \tag{13}$$

$$x_i \sigma_i = x_o \sigma_o \quad \text{(for drug)} \tag{14}$$

$$c_i \sigma_i + R_i \sigma_i + (1 - C)\Omega = c_o \sigma_o + R_o \sigma_o \quad \text{(for moisture)} \tag{15}$$

It is seen from Eq. (13) that solids from the granulating liquid enter into the final product expression and to obtain a final (assayed) content of product (x_o) requires a different input percentage (x_i):

$$x_i = x_o \left(1 + \frac{C\Omega}{\sigma_i}\right) \tag{16}$$

Equation (16) assumes that there are no entrainment losses. With
such losses (of magnitude 2 kg loss per minute with active content
of x_ℓ) Eq. (16) is modified to:

$$x_i = x_o\left(1 + \frac{C\Omega}{\sigma_i}\right) + x_\ell \frac{\ell}{\sigma_i} \tag{17}$$

In general the equilibrium moisture content of a granulation
coincides with the optimum content for compression purposes; it is
also close to the moisture content of incoming raw materials so that
in a practical situation $\sigma_i = \sigma_o$. Using this fact it is possible to
estimate the overall balance between flow rates of granulating liquid
and inlet air by finding σ_o from Eq. (12), σ_i from Eq. (13), and substituting in Eq. (15). This yields:

$$\Omega = \frac{\sigma_i(R_o - R_i)}{1 + C} \tag{18}$$

Operational granulation liquid inlet rates are a function of
air temperature (T °F). Rankell et al. [17] found that for a 30-kg
inventory charge the operational granulating liquid spray rate
(Ω liter/hr) was linearly related to the inlet air temperature by
the equation:

$$\Omega = \frac{T}{20} + 50 \tag{19}$$

Under the same circumstances the mean granule size is a function
of inlet rate of granulating liquid by the relation:

$$\log d = 0.03\Omega + 1.85 \tag{20}$$

where d is the mean "diameter" of the resulting finished granulation.
Equations (19) and (20) are, of course, dependent on equipment and
conditions but the relations:

$$\Omega = \alpha T + \beta \tag{19a}$$

and

$$\log d = \zeta \Omega + \psi \tag{20a}$$

can be expected to hold in general.

Particle-size distributions are, under correct operating conditions, excellent for compression purposes when granulations are made as described above. Particle sizes in natural processes are usually distributed log-normally [18]. In the continuous granulation process the return of fines offsets the distribution slightly, making the amount of fines less than that expected from an unperturbed process; this is shown in Fig. 11 [19].

Spray drying has been attempted for granulation purposes [20] but cost factors, in general, prohibit its use. Flow properties and uniformity of color (in colored granulations) are the primary advantages. Some raw materials used in pharmaceutical manufacturing are spray dried, e.g., spray dried lactose. The material is directly compressible and possesses good flow properties. The former property makes it attractive as a tablet ingredient; it is, however, not recompressible, so that once tablets (or slugs) are made they cannot be reground and recompressed; hence spray dried lactose cannot be used in a slugging operation. It has a tendency to yellow (by formation of hydroxymethylfuraldehyde, as shown by Brownley and Lachman [21]); it is, however, still usable (and used) in many tablet products.

One important pharmaceutical application of a modified spray drying operation is protected vitamin A products. Vitamin A is a highly labile (especially oxygen sensitive) chemical entity and it was not until a satisfactory coating technique was developed that shelf stable vitamin A containing (solid) dosage forms was possible. A problem of coating here results from the fact that the vitamin esters (acetate and palmitate) are liquids. A practical approach,

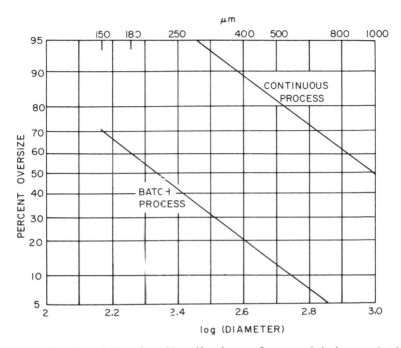

Fig. 11. Particle-size distributions of spray-dried granulations
vis-à-vis tray-dried granulations. The particle-size distributions
are predominantly log-normal (see text). After Rankell et al. [17].
Reproduced, with permission, from J. Pharm. Sci.

developed by Rubin et al. [22] is the emulsification of the vitamin
A ester in a gelatin solution and spraying of the emulsion through
an air space into either a solid catch (modified starch) or a liquid
catch (e.g., mineral oil). The spraying, (by nozzle or rotating disk),
accomplishes formation of spheres of controllable size. The fall
through the airspace accomplishes (a) initial drying and (b) temperature reduction and the catch accomplishes sufficient further drying
and temperature reduction so that the encased vitamin A ester can be
processed further as a solid powder. Particle sizes of 40-80 mesh
are accomplished in this way and the ensuing powder is storage stable
and stable in dosage forms.

III. COATING AND ENCAPSULATION

Spray congealing has found use in several pharmaceutical products. It is frequently desirable to prolong the activity of a dosage form (producing a sustained-release dosage form) and this can be
done by decreasing its dissolution rate; one way of doing this is to
coat it with varying thicknesses of wax, which can be accomplished
by spray congealing. The principle is simply to make a slurry or
solution of the drug in an oil (containing such substances as stearic
acid, castor oil, or glyceryl monostearate in controlled ratios).
Particle size is controlled by the spray. The particles will solid-
ify (congeal) during their fall and can be collected at the bottom
of the dryer. The biological release patterns of the material are
a function of particle size and composition. Glyceryl monostearate
for instance is an emulsifier and enhances the biological erosion of
the spheres (enhances dissolution and hence biological release);
high-melting waxes (castor wax) slow down the release; by proper balance one can therefore achieve a desired biological release pattern.

Scott et al. [23] modified a laboratory spray drier and fed oil
suspensions onto a disk atomizer. They found that the average surface volume diameter d depended on the feed rate Ω, the viscosity of
the feed μ, and the rotational velocity v of the wheel; the experimentally determined relation was

$$d = K\Omega^q v^p \mu^c \tag{21}$$

and typically they found q = 0.17, p = -0.54, and c = -0.02. Raghunathan and Becker [24] found particle-size distributions to be log
normal.

Coating, as mentioned, offers good protection against moisture
and oxygen, and spray congealed solid vitamins (e.g., riboflavin)
are marketed and used extensively in the pharmaceutical industry
(Rocoats, Mercoats); in this application, care is taken to formulate
so that the application of waxes does not slow down biodissolution
and bioavailability.

In tablet coating procedures the solid-gas interphase is again of importance. It is often desirable to coat tablets for protective and aesthetic purposes. Fairly thin, polymeric coats (containing titanium dioxide as an opacifier) are frequently used (e.g., carboxymethyl cellulose, shellac, or cellulose acetate phthalate, the latter for enteric coating). Applications of such films can be made either by a fluid-bed principle, much like the one shown in Fig. 7 (in which case it is referred to as Wurster coating) or the film can be applied in conventional coating pans.

The Wurster coater [25] is shown in Fig. 12. It differs from the fluid-bed drier and granulator by (a) having a set of air-distributing fins (not shown in the figure) for attaining the optimum airflow pattern for fluidization and spray patterns and (b) the fact that coating fluid is applied with the air stream (not against it as in the continuous granulator). The air velocity (v ft/min) required to support a load was found by Wurster [26] to be:

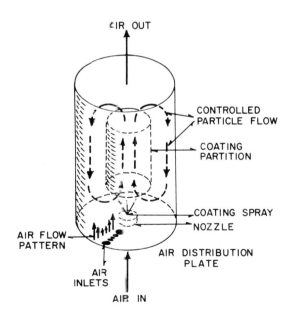

Fig. 12. Wurster coater.

$$v = \kappa\sqrt{m/a} \tag{22}$$

where m is the tablet mass and a its cross section. Wurster found
Eq. (22) to be well obeyed (plotting $\sqrt{m/a}$ versus v) in the sense
that linearity prevailed and the intercept was close to zero. κ
typically has a value of 0.3 $g^{-0.5}$ cm ft min^{-1}.

Loss figures are typically less than 1% in this process and tablet coats can be applied in 25-45 min. This is 2-3 times faster than
by pan coating; the quality of Wurster coated tablets is equal or
superior to that of pan film coated tablets. Scale up is simple
since calculated air velocities, mass build-up, etc. are arithmeti-
cally calculable. Tablets, when exposed to a constant rate of atomized coating solution, should gain weight linearly with time, i.e.,

$$\Delta W = \phi t \tag{23}$$

Wurster [26] found this to be the case in the strictest sense and
typically found $\phi = 1$ mg/hr. Adequate coats may weigh as little as
0.3 mg.

Pan film coating enjoys a high level of popularity; once adequate
film coating solutions are developed it is a simple process: tablets
are placed in a coating pan that is pear shaped and is rotated around
an axis inclined slightly upward. An opening in the face of the pan
allows loading (to about 50% of the volume capacity). Tablets are
tumbled when the pan rotates, and coating can be accomplished by
spraying on coating solution through a suitable spray nozzle. The
spray does not depend on air atomization (airless pumps are used).
The coating solvent is then removed by blowing air on the tablets.
Automated film coating setups have been described by Lachman and
Cooper [27] and Mody et al. [28] and commercially available instal-
lations are being used in the industry today. A fair amount of solvent is involved and for high-volume items, solvent recovery systems
are built into the process [29].

Soft-shell capsules are produced by either the Accogel process
or the R. P. Scherer process. The gas-solid interphase is of importance in both of these processes since the capsules are formed in wet
gelatin, which must subsequently be dried.

Accogel soft-shell capsules are manufactured by casting two films
composed of 20% glycerin, 30% water, and 50% gelatin (approximately)
onto a casting drum, forming a pocket in one, filling it with powder,
and sealing the top film by pressure. The tension in the two films
is equilibrated by tumbling in a coating pan and the capsules are
tray dried at 37°C and 10% relative humidity. The Scherer machine
pumps a suspension of the active ingredients into a pocket formed by
both films, followed by a sealing and tray drying. The processes
have been described in more detail by Carstensen [30].

The drying of the capsules is the subject which is germane to
this text. Many approaches (e.g., extractive solvent removal) have
been tried with poor success and only tray drying seems to be fool-
proof; the problem is the rather narrow moisture limits that must be
maintained in the final product to avoid brittleness (overdrying) or
deformation and tackiness (underdrying). The specifications average
on the order of 20-30 mg of moisture (by USP toluene test [12]) per
g of capsule but, of course, this varies from product to product.

An internal coat (Piccolyte) is applied to the gel film prior
to adding the powder; if the coat is considered impermeable to water
and vapor then the drying is comparable to drying out a cylindrical
slab [10]. This leads to the expression:

$$\ln \frac{W - W_\infty}{W_0 - W_\infty} = \frac{-1}{\kappa} t + \ln \frac{8}{r^2}$$
(24)

κ is the same as in Eq. (9), except that r here denotes film thickness. Data of Accogel soft shell (air fills) is shown in Fig. 13.
Although the initial moisture content of the gelatin film is known
W_∞ will vary from sample to sample. W_∞ can be estimated, however, by

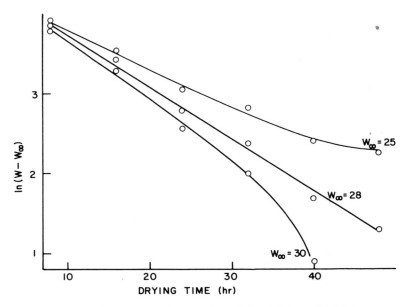

Fig. 13. Drying curve of capsules (air fills) of thickness
0.0225 in. Top curve is obtained when W_∞ = 25, middle when W_∞ = 28,
bottom when W_∞ = 30 mg of moisture.

iteration, such as is shown in Fig. 13. The equation for the line
corresponding to W_∞ = 28 mg is

$$\ln \frac{W - 28}{82} = -0.0641t - 0.633 \tag{25}$$

The intercept should be $\ln(8/\pi^2)$ = -0.213, and hence complete correlation with the theoretical intercept is lacking, although order of
magnitude is correct. This is presumably due to the fact that Eq.
(24) is derived for an infinite slab, whereas Eq. (25) very much applies to one of finite length. Estimates of diffusion coefficients
can be gotten from intercept and slope: if intercept is denoted β
and slope α, then κ = $-1/\alpha$ = $r^2/(\pi^2 D)$; from the data in Fig. 13, it
is seen that D = 2.6 × 10^{-7} cm /sec, which is of the expected order
of magnitude.

IV. STERILIZATION AND HANDLING

Another pharmaceutical case where gas-solid interphases are of importance is in ethylene oxide sterilization. Many parenteral products are chemically unstable in solution and potencies of such solutions might be satisfactory from a therapeutic point of view for a day or two, but not for a sufficient time to allow the solution to be marketed. To make a usable product is is common practice to market the drug as a powder in the vial. The powder is sterile and prior to use, it is reconstituted with physiological saline solution (or other prescribed diluent). By syringing this into the vial a solution is formed which can be injected.

One of the problems in this application is rendering the powder sterile; if the drug can stand high heat (170°C for 1 hr) in the solid state then it can be heat sterilized. Frequently this is not possible and in such cases the powder may be ethylene-oxide sterilized. In this process the vials are filled with the appropriate amount of solid powder (or powder mix) and placed uncapped in an autoclave. The autoclave is evacuated and sterilization gas is let in (usually via an expansion tank). The sterilization gas is 10% ethylene oxide and 90% carbon dioxide [31] and is commercially available as Carboxide in 30- or 60-lb containers. The dilution is necessary because ethylene oxide is highly explosive. After exposure of the powder (e.g., for 1 hr) the autoclave is evacuated and purged twice with sterile nitrogen (or air). Vials are then aseptically plugged (in certain arrangements directly in the autoclave).

Ethylene oxide sterilization has been very popular but has in recent years been under scrutiny by the Food and Drug Administration. Satas [32] has shown that (a) ethylene oxide sorbs strongly onto many materials and (b) it is catalytically converted to ethylene glycol by many materials when trace moisture is present. Ethylene oxide is very reactive so that if evacuation and purging does not remove it all, it is liable to give rise to unwanted reactions on storage in the vial; it (as well as ethylene glycol) is toxic. Ethylene glycol,

if formed during sterilization, is difficult to remove by purging because of its low volatility. When ethylene oxide is used in present
day pharmaceutical products and operations, model experiments must
be performed to show absence of ethylene oxide and its degradation
products in the final marketed vial.

Recently general operational procedures in the pharmaceutical industry have come under scrutiny from the Food and Drug Administration;
this was partially initiated by certain cross-contaminations, e.g.,
penicillin showing up in trace amounts in tablets made on other nearby
equipment such as tablet machines. The problem was not one of clean-
up (i.e., of subsequent batches on the same equipment) but one of airborne contamination (simultaneous batches on different equipment) in
most cases. Stringent rules relating to scheduling, and in some cases
isolation of production facilities for certain products minimize such
occurrences. The regulatory requirements in this respect are spelled
out in the Federal Register under "Good Manufacturing Practice," Sec.
133.3.

Requirements for oral products are not as stringent as for in-
jectables. Fischer [33] suggest the following for nonaseptic products: (1) For storage (drugs, excipients, containers, liquids):
natural cross-ventilation. (2) For dispensing (weighing out specific
amounts of specific substances for a batch of a product): 4-in i.d.
(internal diameter) flexible metal hose to bulk drum, 3-in i.d. flexible metal hose to weighed container, both exhausting at 120 cfm.
The hourly air renewal rate should be about 15. (3) For granulation:
areas should have an hourly air renewal rate of 10-15 and are usually
not air conditioned. (4) For tableting: areas require 30-50% relative humidity; strongly colored tablets (e.g., tablets containing pyr-
idium) should be made in separate areas with separate air supply.
(5) For hard-shell encapsulation: 40-55% relative humidity required.

Parenteral powders that are to be sterilized later require clean
rooms; particulate matter that may find its way into the vials may

still allow for a sterile product, but the solution will not be clear
subsequent to sterilization; furthermore the chance of pyrogenicity
is larger with particulate contamination, and the short discussion to
follow will therefore center on some of the principles of clean rooms.

There is no general rule for air supply planning other than the
pressure rule to be mentioned shortly. In present-day layouts, laminar flow arrangements are used much more frequently than previously;
an example is shown schematically in Fig. 14a in comparison with the
conventional overhead arrangement in Fig. 14b. The sequence for air-
ducts is one of (A) coarse filter, (B) blower, (C) fine filter, with
the latter as close as possible to the room being supplied. The air
renewal rate is 10-20 in a conventional setup, ten times as large
in a laminar setup. In spite of the apparent advantage to laminar
setups there are several drawbacks: (a) operators are a source of
contamination, (b) strict rules regarding flow of materials, movement,
etc. must be observed in order to minimize turbulence; these rules may
be (and most often are) contrary to optimum operational procedures.
It is, therefore, typical to confine laminar flow areas to the aseptic
point, e.g., in form of laminar flow hoods in which aseptic filling
(or other aseptic operations) takes place, i.e., to combine the advantages of overhead flow with those of laminar flow (Fig. 14b).

The pressure gradients are always maintained so that the "cleanest" area has the higher pressure; in that manner opening a door will
cause air to flow into the less clean room, hence the flow of contami-
nants is always away from the cleaner area. Fischer [33] suggests the
following for a four-step ampule cleaning and filling setup (Fig. 14c):
a hallway (A) leads into a room (B) where ampules are removed from
boxes; this leads into the wash kitchen (C) equipped with autoclaves;
the autoclaves can be discharged from the rear in room (D) where filling takes place aseptically. D is often connected to a robing room
(E). D is obviously the cleanest room and has the highest pressure
(3-4 mm H_2O). Pressures then fall off in either direction and are

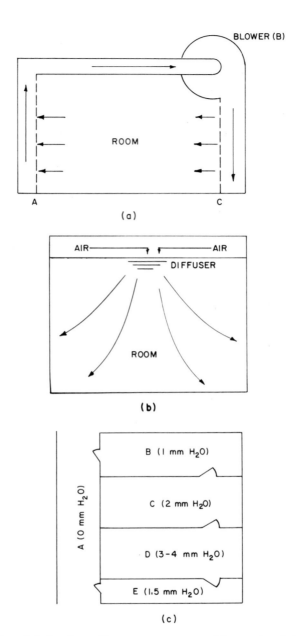

Fig. 14. (a) Laminar flow arrangement; (b) overhead flow (with laminar flow hoods) room; (c) arrangement of rooms with highest pressure in the "cleanest" room: corridor (atmospheric) (A), unpacking (1 mm H_2O over atmospheric) (B), washing (2 mm over atmospheric) (C), and sterile filling (3-4 mm H_2O over atmospheric) (D).

zero in the hallway. Many other layouts are, of course, possible.
As a rule planning should start from the "cleanest" area (highest
pressure, e.g., 3-4 mm H_2O) and branch out to reach (last) the generally accessible, least clean area (having atmospheric pressure). A
thorough, detailed review of these problems has been published by
Fischer [33].

REFERENCES

1. R. Carr, Chem. Eng., 67, No. 4, 121 (1960).

2. G. Gold, R. N. Duvall, B. T. Palermo, and J. G. Slater, J. Pharm. Sci., 57, 667, 2153 (1968).

3. T. M. Jones and N. Pilpel, J. Pharm. Pharmacol., 18, 81 (1966).

4. J. T. Carstensen, Theory of Pharmaceutical Systems, II - Hetero- geneous Systems, Academic Press, New York, 1973, p. 227.

5. W. E. Garner, Chemistry of the Solid State, Academic Press, New York, 1955, p. 213.

6. C. Shepherd, C. Hadlock, and R. Brewer, Ind. Eng. Chem., 30, 388 (1938).

7. T. R. Oliver and D. M. Newitt, Trans. Inst. Chem. Eng., 27, 1 (1949).

8. J. F. Pearse, T. R. Oliver, and D. M. Newitt, Trans. Inst. Chem. Eng., 27, 9 (1949).

9. K. Ridgway and J. A. B. Callow, J. Pharm. Pharmacol., 19, 155S (1967).

10. W. Jost, Diffusion, Academic Press, New York, 1952, p. 46.

11. C. Pitkin and J. T. Carstensen, J. Pharm. Sci., 62, 1215 (1973).

12. United States Pharmacopeia, 17th Revision, Mack Printing Co., Easton, Pa., 1955, p. 925.

13. Liquids-Solids Blender, The P-K Company, East Stroudsberg, Pa.

14. L. Lachman, U. S. Patent, 2,877,159, 3/10, 1959.

15. D. E. Wurster and G. P. Polli, J. Pharm. Sci., 50, 403 (1961).

16. M. W. Scott, H. A. Liberman, A. S. Rankell, F. S. Chow, and G. W. Johnston, J. Pharm. Sci., 52, 284 (1963).

17. A. S. Rankell, M. W. Scott, H. A. Liberman, F. S. Chow, and J. V. Battista, J. Pharm. Sci., 53, 320 (1964).

18. J. T. Carstensen and M. N. Musa, J. Pharm. Sci., 61, 223 (1972).

19. M. W. Scott, H. A. Liberman, A. S. Rankell, and J. V. Battista, J. Pharm. Sci., 53, 314 (1964).

20. A. M. Raff, M. J. Robinson, and E. V. Sverdres, J. Pharm. Sci., 50, 76 (1961).

21. C. Brownley and L. Lachman, J. Pharm. Sci., 53, 452 (1964).

22. S. H. Rubin and E. DeRitter in Vitamins and Hormones (L. W. Marson, ed.), Academic Press, New York, 1953.

23. M. W. Scott, M. J. Robinson, J. F. Pauls, and R. J. Lantz, J. Pharm. Sci., 53, 670 (1964).

24. Y. Raghunathan and C. H. Becker, J. Pharm. Sci., 57, 1748 (1968).

25. D. E. Wurster, U.S. Patents 2,648,609 and 2,799,241 (1959).

26. D. E. Wurster, J. Am. Pharm. Assoc., Sci. ed., 48, 451 (1959).

27. L. Lachman and J. Cooper, J. Pharm. Sci., 52, 490 (1963).

28. D. S. Mody, M. W. Scott, and H. A. Lieberman, J. Pharm. Sci., 53, 949 (1964).

29. J. W. Drew, Chem. Eng. Progress, 71 (2), 93 (1975).

30. J. T. Carstensen: Theory of Pharmaceutical Systems, II - Heterogeneous Systems, Academic Press, New York, 1973, pp. 282-285.

31. C. R. Phillips and S. Kaye, Am. J. Hyg., 50, 270 (1949).

32. D. Satas, J. Pharm. Sci., 53, 675 (1964).

33. B. Fischer, J. Mond. Pharm., 14, No. 1-2, 108 (1971).

Chapter 7

G. A. Hohner

Research and Development
The Quaker Oats Co.
Barrington, Illinois

I. INTRODUCTION

In Chapters 1-4 the principles of solids-handling systems were discussed in terms of the properties and mechanics of materials in general. In this chapter the unique properties of solid foodstuffs and the special considerations necessary for the design of handling and storage systems for foods are discussed.

The quantity of solid foodstuffs stored, conveyed, transported, and processed each year is enormous. These foodstuffs include raw materials, processed ingredients, byproducts, and finished products. Excluding frozen products, solid foodstuffs are largely low or intermediate moisture-content products. They appear in a wide range of physical forms from dice and granules to powders, many of which are handled in bulk. Discussions in this chapter are limited primarily to these foodstuffs.

The principles discussed previously are largely applicable to the design and operation of solids-handling systems for foodstuffs, provided certain additional requirements are observed. These requirements are related to the biological nature of foodstuffs.

Bulk, solid foodstuffs cover a wide range of values with respect to fundamental physical properties of materials. In recent years methods have been developed or adapted from other branches of materials science to obtain reliable values for physical parameters of foodstuffs. Mohsenin [1] has presented a review of the subject of physical properties of biological materials including significant amounts of data. In spite of this and other sources, many products exist for which few data suitable for use in design are available. The designer of processes or systems for handling foodstuffs must often acquire his own experimental data on the specific material in question.

Successful storage and handling of solid foodstuffs require an understanding of the biological nature of the product. Solid foodstuffs are not inert; they react and interact with the local environment. They adsorb and desorb water vapor in response to changes in relative humidity and temperature. As a result the physical parameters also become dependent on environmental variables.

Major importance must be placed on understanding and achieving product stability during long-term storage. Sorption phenomena of biological materials play a key role in product stability. These phenomena and some recent theories to explain them are discussed to provide a basis for understanding the relationships between the physical-chemical state of the biological product relative to the environment and how this affects storage stability. Since limitation of the chemical potential or activity of water is the primary means of preserving solid foodstuffs, it is not surprising that the degradation mechanisms of stored foodstuffs relate to the quantity and state of water in the product.

All biological materials, including foodstuffs, support the growth of microorganisms under certain environmental conditions. The rate of growth of a microbial population in a food product depends on the chemical and physical composition of the product, environmental conditions, and requirements of the organism. The maintenance of microbial stability in storage and handling systems for foodstuffs involves a public health responsibility of the first magnitude. This responsibility must be largely borne by the designer of the system. Major means of controlling microbial populations in solid foodstuffs will be discussed.

Besides microbial spoilage, product stability may be threatened by various biochemical reactions that can reduce the nutritional value and organoleptic acceptance of foodstuffs during storage. The nature and rate of these biochemical reactions are dependent on the environmental conditions present or created during storage. The effect of the various degradation reactions and their link to the moisture equilibrium condition of the product will be considered later in this chapter.

The design of bulk storage bins, hoppers, and openings is considered in the last section. The methods of Jenike [2,3] are discussed and applied to foodstuffs. Of primary importance in the design of bulk storage facilities for foodstuffs are conditions that insure a first-in, first-out flow regime. Such a design guarantees that no product remains in a dead spot in a storage bin indefinitely while being bypassed by newer, fresher product. Foodstuffs trapped in areas of no-flow for indefinite periods of time are subject to various spoilage mechanisms discussed in the following sections.

II. PRODUCT STABILITY

Solid foodstuffs are unique in that they are not inert during storage. Thus, they differ from many other solids that are handled

in the process industries. Unfortunately, most changes that occur
during storage and handling are negative in nature, tending to result
in a less acceptable product. Exceptions to this are processes such
as the aging of cheese to develop typical flavors. Dynamic changes
in foodstuffs are the result of complex biochemical chain reactions
that are not completely understood. However, sufficient knowledge is
available to say with certainty that the effect of degradation mechanisms can be minimized by proper storage conditions.

The role of product moisture in maintaining stability during
storage is discussed in this section. Product moisture controls the
extent of microbial growth and spoilage in solid foodstuffs and has
an important bearing on the type and rate of several biochemical reactions, which must be controlled to maximize product storage life.

A.   THE INFLUENCE OF PRODUCT MOISTURE

Virtually all foodstuffs adsorb and desorb water vapor in moving
toward a state of moisture equilibrium with the local environment;
they are hygroscopic. The moisture content toward which a product
equilibrates in a given situation is dependent on the available surface area per unit weight of the foodstuff and the number and type of
reactive sites present on the surface. Surface area, in this context,
refers to all intercellular spaces within the product in addition to
exposed surfaces. Adsorbed moisture on the surfaces of a solid foodstuff during storage and handling can have a profound effect on the
biochemical and microbiological stability of the product.

The amount of water vapor adsorbed is dependent on the water
activity of the local atmosphere and the temperature. The amount adsorbed to a solid foodstuff in equilibrium with a particular environmental condition is commonly called the equilibrium moisture content.
When equilibrium moisture content is plotted versus the water activity
of the environment for a particular temperature the plots are known
as equilibrium moisture isotherms. Figure 1 shows isotherms for
freeze-dried beef slices, dehydrated carrots, and whole corn as examples of the sigmoidal-shaped isotherms typical of biological mate-

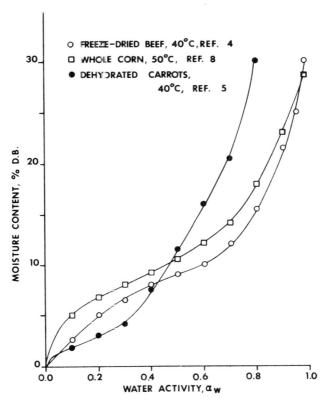

Fig. 1. Sigmoidal shaped, equilibrium moisture isotherms typical of biological materials.

rials. Equivalently, equilibrium moisture content may be plotted versus atmospheric relative humidity. For the dilute system of water vapor in air the following equality is valid:

$$a_w = P/P_o \tag{1}$$

$(P/P_o) \times 100 = $ relative humidity, %

The effect of temperature on the equilibrium moisture content is shown in Fig. 2 for cornstarch. Not surprisingly, the equilibrium

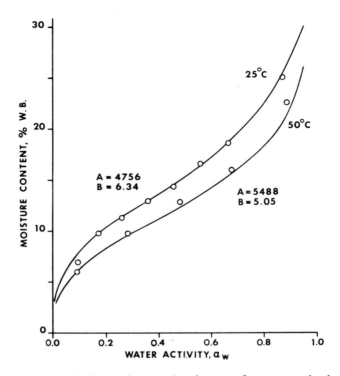

Fig. 2. Equilibrium moisture isotherms of cornstarch showing the effect of temperature.

moisture content decreases with increasing temperature at a given water activity. The implications of this phenomenon on the design of long-term storage facilities for solid foodstuffs will be discussed later.

A mathematical explanation of the process of adsorption of water vapor on a porous solid has been the subject of many investigations. A brief explanation of a few proposed mechanisms of sorption is of considerable assistance in understanding or predicting storage stability results in foodstuffs.

Langmuir [6] was apparently first to attempt to explain adsorption of vapors to solids in a theoretical manner. The Langmuir isotherm equation is applicable only to adsorption in a monomolecular

layer on a solid surface. The equation is not applicable to foodstuffs that clearly show multilayer adsorption.

In 1938, Brunauer et al. [7] proposed the Brunauer, Emmett and
Teller theory of multilayer adsorption of gases on solids. Equation
(2) is the Brunauer, Emmett and Teller equation of an equilibrium moisture isotherm:

$$\frac{P}{M(P_o - P)} = \frac{1}{M_m C} + \frac{C - 1}{M_m C} \frac{P}{P_o} \tag{2}$$

where M is the moisture content, dry basis; C is a constant related
to the heat of sorption, and P is the water vapor pressure in equilibrium with moisture content, M. The Brunauer, Emmett and Teller
theory does not fit the equilibrium moisture isotherms of food products over the entire water activity range. However it is accurate
at low moisture levels and is often used to predict the moisture content equivalent of a monomolecular layer of water molecules adsorbed
over the entire surface, M_m. The monomolecular moisture level can be
calculated from Eq. (2) by plotting $P/M(P_o - P)$ versus P/P_o. The y
intercept is $1/M_m C$ and the slope is $(C - 1)/M_m C$. Simultaneous solution of

$$\frac{1}{M_m C} = \text{intercept} \quad \text{and} \quad \frac{C - 1}{M_m C} = \text{slope} \tag{3}$$

gives both C and M_m.

The calculated monolayer may not in fact represent a continuous
film of adsorbed water vapor, but it does correspond to water molecules being adsorbed to all available reactive adsorption sites on all
available surfaces of the product. The adsorbed monolayer of water
vapor acts to suppress the reactivity of the sites on the product surface. The implications that the protective monolayer has for maintenance of product stability are discussed in Sec. II-C.

Chung and Pfost [8], 1967, presented a derivation of equilibrium
moisture for cereal grains and commercial products made from cereals.

They assumed that sorptive sites on the product surfaces exert strong
attractive forces on the nearby water vapor. The forces are assumed
to be sufficiently great to cause many layers to be adsorbed on the
surface. Each layer is assumed to be under compression from the
layers adsorbed over it.

The authors assumed that the free energy of the adsorbed moisture
decreases exponentially with increasing thickness of the adsorbed
layer and that the free energy function of water vapor is temperature
dependent. From these assumptions Eq. 4 was derived for adsorbed
moisture as a function of atmospheric water activity and temperature.

$$M = -B \ln \left(\frac{-RT}{A} \ln a_w \right) \tag{4}$$

where A and B are temperature-dependent constants which must be evaluated from experimental isotherm data. Accuracy of the derived equation in fitting experimental isotherms of cereal products is demonstrated in Fig. 2.

Finally, Ngoddy and Bakker-Arkema [9] recently published a generalized theory of sorption phenomena in biological materials in which
they combined fundamental concepts of previous theory. Their objective was a mathematical expression that could be used to predict the
equilibrium moisture isotherms of biological products.

The generalized theory is based largely on multiple-layer molecular adsorption theory combined with capillary condensation theory.
Pores and capillaries of the porous material are hypothesized to fill
with water by two mechanisms: molecular adsorption and capillary condensation. Water vapor is adsorbed to reactive sites on the capillary
walls. The adsorbed water vapor reduces the effective radius of the
capillary or pore. Condensation of water vapor may be then spontaneously induced in a capillary by the reduced radius of curvature. The
equilibrium vapor pressure of a liquid in a cylindrical capillary is
reduced below the saturation point by surface tension in the curved
liquid surface. The vapor pressure of the adsorbed liquid is related
to the radius of curvature of the adsorbed surface by the familiar
Kelvin equation

$$r_c = \frac{2\gamma V}{RT} \ln \frac{P_o}{P} \tag{5}$$

where r_c is the radius of curvature, γ is the surface tension of the adsorbed liquid, V is the molar volume, and R is the universal gas constant. So long as the radius of curvature continues to decline, condensation of more vapor will continue.

The authors derived Eq. 6 as the equilibrium moisture isotherm equation for a biological solid in which the distribution of pore radii is characterized by the power-law relationship of Eq. (7).

$$M = \rho \frac{\xi}{\eta}(Z^{\eta} - \lambda^{\eta}) \tag{6}$$

where

$$Z = 4.24(\ln X_1)^{-1/2} + \frac{\gamma V}{RT \ln X_1}$$

$$\lambda = 4.24(\ln X_2)^{-1/2} + \frac{\gamma V}{RT \ln X_2}$$

$$X_1 = \frac{P_o + P_m}{P + P_m}$$

$$X_2 = \frac{P_o + P_m}{P_m}$$

and where ρ is the density of the adsorbed phase, and ξ and η are parameters of the pore radii distribution $D(r)$.

$$D(r) = \frac{3\xi r^{\eta-4}}{4\pi} \tag{7}$$

Theoretically, the pore radii of a solid foodstuff could be characterized sufficiently by evaluating ξ and η from one equilibrium moisture isotherm or by more direct measurements. The amount of moisture in equilibrium with any particular vapor pressure and temperature could then be predicted from Eq. (6). In practice, it was found that the pore-structure parameters ξ and η were temperature dependent and not constant. This may be due to actual thermal-induced changes in the pore structure during sorption or because of an inadequate model. Also, P_m, which was initially defined as vapor pressure corresponding to the monomolecular moisture condition, may have lost its physical significance due to a change of axes to obtain Eq. (6) in final form.

In summary, Eq. (6) proves to be a highly accurate three-parameter model for the equilibrium moisture isotherms of biological products; however, the utility of the equation may be hampered by its complexity. Nevertheless, a significant contribution to the understanding of the complex process of sorption in biological products was made by its logical derivation.

The accuracy of Eq. (6) is demonstrated for isotherms of freeze-dried beef and whole corn in Fig. 1. The theoretical models indicate the probable mechanisms of water sorption in biological products and provide a frame of reference to explain observed microbial and biochemical stability phenomena.

Solid foodstuffs often exhibit hysteresis in the amount of water sorbed, depending on whether water is being adsorbed or desorbed. That is, the processes of adsorption and desorption do not follow the same path on the equilibrium moisture isotherm in response to changes in water activity. Typically the desorption curve lies above the adsorption curve.

It has often been observed and reported that the hysteresis loop tends to disappear as repeated cycles of sorption and desorption are impressed upon the material. From a practical point of view, however, it is unlikely that hysteresis is important during normal cycles of adsorption and desorption where neither total hydration nor total dehydration are ever attained. The loci of equilibrium states probably lies along the desorption isotherm for materials that have never been totally dehydrated.

Another important facet of sorption in biological materials is the energy required to desorb water vapor from the solid. This energy requirement is substantially greater than the free-water heat

of vaporization and varies with the amount of moisture sorbed. At
higher moisture contents the energy required to desorb water vapor
that is weakly bound is less than for tightly bound water vapor at
lower moisture contents. The isoteric heat of sorption, ΔH, for adsorbed water vapor can be expressed as

$$\frac{\partial(\ln P)}{\partial(1/T)} = \frac{-\Delta H}{R} \tag{8}$$

Assuming that ΔH is invariant with temperature it can be evaluated
from two separate experimental desorption isotherms by a rearrange-
ment of Eq. (8)

$$\Delta H = R\left[\frac{T_1 T_2}{T_2 - T_1}\right] \ln\frac{P_2}{P_1} \tag{9}$$

where P_1 and P_2 are the equilibrium vapor pressures of water at
temperatures T_1 and T_2 respectively.

Figure 3 shows the heat of desorption for various fractions of
the commercial wet corn milling process as a function of their moisture content. The energy required for drying solid foodstuffs to
low moisture content is substantially greater than for the evaporation of free water, and highly dependent on the composition of the
material. Cornstarch is a nearly pure polymer of glucose with primarily $\alpha,1$-4 linkage. In raw starch the polymer is tightly coiled
in a granular structure. From Fig. 3 cornstarch obviously has the
strongest bonding energy to adsorbed water vapor of the various corn
fractions considered. The desorption energy for water from cornstarch at 4% moisture content is more than 50% greater than the free
water energy of vaporization

Corn hull and gluten have progressively less carbohydrate and
more protein, fiber, and lipid components. At the same moisture content their desorption energy of water vapor declines, approximately

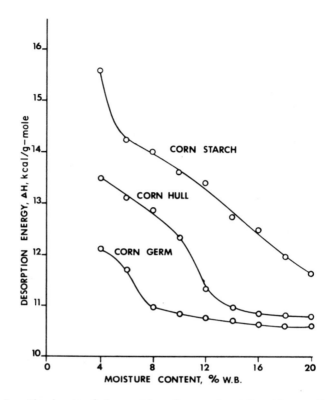

Fig. 3. The heat of desorption for various fractions of corn.

proportional to the lipid concentration. Corn germ has the highest
lipid content of any fraction and the lowest desorption energy. The
implications that the quantity and state of product moisture have
for the microbial and biochemical stability of foodstuffs during
storage and handling are discussed in the following sections.

B. MICROBIAL CONSIDERATIONS

     Virtually all foodstuffs readily support growth of numerous
species of microorganisms when suitable environmental conditions
exist for microbial growth over sustained periods of time. In fact,

many food processes such as manufacture of cheese, wine, and bread depend on controlled growth of specific species of microorganisms for achieving a desired effect in the final product. The normal situation facing the designer of a system handling solid foodstuffs is the necessity of controlling or eliminating growth of many, usually unidentified, species of microorganisms that might naturally exist in the product or become introduced into the system from a foreign source.

In foodstuffs three general groups of microorganisms must be considered: molds, yeasts, and bacteria. Each group has relatively specific microenvironmental requirements necessary for stimulating growth and reproduction. In general, it may be assumed that microorganisms are not seriously limited with respect to nutrients when foodstuffs are the substrate. Other than the quantity of nutrients available to the organism, important environmental conditions determining the rate of microbial growth are temperature, pH, water activity, osmotic pressure, oxygen concentration, and the concentration of various natural or foreign inhibitors to which the organism is exposed.

Most raw foodstuffs contain a natural surface flora that may be composed of many species. Quantities of organisms present may vary from a few hundred to several million per gram for foodstuffs that originate in or on the soil. As bulk, raw foodstuffs are stored, handled, processed, and moved along commercial channels of the food industry, the quantity and type of organisms present may change. Highly processed food ingredients may have fewer than 1,000 total organisms per gram, while dry spices imported from tropical climates may have several million.

Growth and reproduction of pathogenic organisms in foodstuffs can, under certain circumstances, result in production of sufficient toxin to cause illness or death in humans if the organisms or their toxins are consumed without destruction. Much more commonly, large concentrations of organisms result in sufficient putrefactive odor, gas, or acid production to make consumption of the product to be un-

likely. Large organism populations may still cause sufficient spoilage to reduce the economic value of foodstuffs due to unsatisfactory
flavor, odor, or appearance.

1. The Influence of Water Activity

Preservation of solid foodstuffs from microbial attack during
storage is based on removal of one or more of the essential conditions for microbial growth, and perhaps, in addition, on reduction
of the existing population by pasteurization or sterilization treatment. Solid foodstuffs are, by definition, of low or intermediate
moisture content. Therefore, microbial growth in solid foodstuffs
is generally limited by the availability of free water in the product.
Water that is not adsorbed to the biological solid but is present
largely as capillary condensate is often called free water and is
available to microorganisms. As discussed in the previous section,
moisture content in biological material can be described in terms of
the water activity and temperature of the environment that is in
moisture equilibrium with the foodstuff.

In general, bacteria require more free moisture than yeasts and
yeasts more than molds. Mossel and Ingram [10] list the lower limits
of water activity that will permit the growth of spoilage organisms.
See Table 1.

TABLE 1

Limiting Values of a_w Which Support Growth

Type of organism	a_w
Normal bacteria	0.90
Normal yeast	0.88
Normal molds	0.80
Halophilic bacteria	0.75
Osmophilic yeasts	0.60

Spoilage by mold growth is more likely than by yeast or bacteria in solid foods. It is generally assumed that moisture contents in solid foodstuffs that are in equilibrium with water activity of 0.70 or less will not support growth of microorganisms. This value establishes the upper limit of the desirable storage range for solid foodstuffs on the water activity scale. Ways in which this range is extended to higher values are discussed in the following section.

In Fig. 2 attention was called to the fact that equilibrium moisture content of a biological material declines with increasing temperature at constant vapor pressure. This phenomenon becomes of practical significance with respect to microbial growth in stored foodstuffs. For example, if a cereal grain is placed in bulk storage in fall or winter months a low temperatures it can be safely stored with a moisture content above 20%. With warmer temperatures in the spring, or if the product is transported to a warmer climate, the equilibrium point moves to a higher temperature isotherm, but at the original moisture content. This change results in movement to a higher equilibrium water activity, which may be above the limiting value for microbial growth, resulting in spoilage. The expected microbial growth would be largely molds, since these microorganisms have the lowest limiting water activity value.

2. Kinetics of Microbial Growth and Death

Microbial populations typically pass through definable stages of growth, based on the amount of time that elapses after favorable conditions for growth have been established. Initially there is a lag phase when there is little or no increase in population. This is followed by an exponential phase in which the rate of reproduction is rapid and constant. Finally the population reaches the sustainable limit for the local conditions, called the maximal stationary phase.

In storage and handling systems it is desirable to extend the lag phase as long as possible. More specifically, it is important to know the microbial state at which a population will equilibrate in a continuous process. If a process location has conditions that can support microbial growth, it will constantly innoculate all subsequent, or downstream, locations even though they are initially sterile or well removed from normal growth conditions. The concentration of organisms in such a location can continue to increase and such a condition is aggravated by dead spots, low flow rates, backmixing, and other deviations from plug-flow conditions.

Short experimental or pilot plant trials on a nonaseptic process or conveying system may not detect locations of potential microbial growth due to the lag phase of the population growth curve, which must elapse before significant numbers of organisms are detected. It is important to know that a process is at microbial equilibrium before drawing conclusions about the long term microbial stability of the process.

Processes that are used to reduce existing microbial populations in foodstuffs are often classified as either sterilization or pasteurization. Either may be accomplished by moist heat, dry heat, certain chemicals, and various high-energy sources such as ionizing radiation. In the food industry moist heat is by far the mechanism of choice. Moist heat is much more lethal to microbial populations than dry heat, including superheated steam. Chemical sterilization will be discussed in the next section (II-B-3).

Differences between sterilization and pasteurization are in degree only. Sterilization is intended to denature or kill all microorganisms in a given space. The practical limit of sterility, often referred to as "commercial sterility," is defined as a residual microbial population incapable of reproduction under hermetically sealed conditions. Thermal sterilization of foodstuffs is commonly conducted at temperatures in the range of 230-300°F. In thermal sterilization design, the effect of the process on spores of thermophilic, spore forming species of bacteria is used as a standard since these microorganisms are the most thermal-resistant species.

Pasteurization is much less severe and is usually intended to kill only molds, yeast, and vegetative cells of bacteria. Pasteurization does not inactivate bacterial spores. It is conducted at temperatures from 140-210°F.

Many lethal processes used for sterilization or pasteurization of foodstuffs have been observed to give a first-order decline in a microbial population with time. That is,

$$\ln(N/N_o) = -kt \tag{10}$$

where N = number of viable organisms per unit volume, N_o = original number of viable organisms, k = rate coefficient, and t = time elapsed.

Stumbo [11] has discussed thermobacteriology as it relates to foods. In food processing, by convention, the rate of microbial death for a certain species at a given temperature is measured in terms of the decimal reduction time, D,

$$D = 2.303/k \tag{11}$$

A second sterilization process parameter z measures the temperature dependence of D

$$z = \frac{(T_2 - T_1)}{\log(D_2/D_1)} \tag{12}$$

A typical value of z for thermophilic spores in the presence of lethal temperatures of moist heat is 18 F. Typical values for D at 250°F are given in Table 2 for various spoilage organisms. Other lethal mechanisms such as toxic chemicals have far different sterilization process parameter values.

TABLE 2

Typical Values of the Parameter D at 250°F

Organism	D value
Thermophiles	
Bacillus stearothermophilus	4.0 - 5.0
Clostridium thermosaccharolyticum	3.0 - 4.0
Clostridium nigrificans	2.0 - 3.0
Mesophiles	
Clostridium botulinum	0.1 - 0.2
Clostridium sporogenes	0.1 - 1.5

The lethality of a thermal process is measured as the integrated effect of a lethal condition over a given time. For moist-heat sterilization in the food industry lethality is defined in terms of F values. One unit is the lethality accumulated at 250°F for 1 min. It can be expressed mathematically as

$$F = \int_0^t 10^{(T-250)/z} \, dt \tag{13}$$

where T is the temperature of the point in question, in degrees Farenheit and t is time in minutes. Obviously this definition and concept of lethality can be extended to any type of lethal process which follows the first-order kinetics of microbial death described above.

Procedures for sterilization and pasteurization of liquid foods with moist heat are well developed. However, these techniques are not easily applied to solid foodstuffs of low or intermediate moisture content that cannot be hydrated without damage to the product. Fabricated foods can easily be thermally processed while in the liquid or slurry state. For products such as dry spices, alternate means of sterilization are clearly needed.

3. Gas Sterilization

Thermal sterilization is by far the most common means of reducing or eliminating microbial populations in foodstuffs. However,
for some solid foodstuffs, which could be damaged by exposure to
sterilizing temperatures, sterilization by exposure to toxic gases
may be an attractive procedure. Gaseous sterilization is also applicable to storage and handling facilities, and to processing equipment.

The chief advantage of gaseous sterilization is the capability
of sterilization at ambient or only slightly elevated temperatures.
Gas sterilization is inherently a surface phenomenon; however, by
appropriate technique of application the gas can adequately penetrate
into bulk foodstuffs and into the internal surfaces of porous food
products. Gas sterilization is convenient and economical, although
some techniques in utilizing gas involve cyclic pressures and require
a pressure chamber [12]. The chief disadvantages are damage that
these highly reactive gases may cause in foodstuffs, the problem of
possible human intoxication during application of sterilizing gases,
and the problem of the residual toxicity that may be difficult to
remove, depending on the product.

Numerous gases have been used as sterilizing agents in food,
pharmaceuticals, and various medical products. These include ethylene
oxide, propylene oxide, sulfur dioxide, formaldehyde, ozone, and
chlorine. Ethylene oxide and, more recently, propylene oxide have
had significant use in solid foodstuffs, especially herbs and spices
in bulk containers. Subsequent discussions in this section will be
limited to ethylene oxide and propylene oxide.

Ethylene oxide, $CH_2(O)CH_2$, has a boiling point of 51.4°F at 1
atm pressure. It is flammable and explosive in volumetric concentrations in air from 3.6 to 100%. Ethylene oxide is purchased commercially as a mixture with carbon dioxide or chlorofluorhydrocarbons
in sufficient concentration to make the mixture nonflammable. The
gas is toxic to humans to about the same degree as ammonia but is
more dangerous because of a pleasant aroma.

Use of ethylene oxide for sterilization apparently causes tex-
ture or flavor damage in few foodstuffs; however, nutritional damage
has been reported. Bakerman et al. [13] have reported destruction
of 40% of thiamine and lesser destruction of other vitamins in a rat
diet exposed to ethylene oxide for 18 hr at room temperature. This
is an extremely long-time, low-temperature sterilization. Little is
known about the kinetics of vitamin destruction versus the kinetics
of bacterial death in toxic gas sterilization; however, in the case of
sterilization with heat it is possible to design processes lethal to
bacteria which cause minimal vitamin destruction. The same may be
true of chemical sterilization when complete kinetic information is
available. For example, exposure to higher temperatures for far
shorter times may provide the same bacterial destruction with less
vitamin damage than that measured by Bakerman et al. [13].

Propylene oxide, $CH_3CH_2(O)CH_2$, has similar properties to ethylene oxide. It has a higher boiling point (63°F) and is less flammable, less effective as a sterilizing agent, and less toxic to
humans. Flammability limits of propylene oxide are 2.1 to 21.5% by
volume in air, and generally no dilution with inert gases is required.

Propylene oxide appears to be about one-tenth as biologically
active as ethylene oxide. In the presence of sufficient moisture,
propylene oxide hydrolyzes slowly to propylene glycol, which is harm-
less. Ethylene oxide, on the other hand, forms toxic ethylene glycol on hydrolysis. As a result, ethylene oxide has been largely replaced by propylene oxide in accordance with the 1958 Food Additive
Amendment to the Food, Drug, and Cosmetic Act.

Numerous investigators, including Michael and Stumbo [14] have
confirmed that the mechanism of bacterial death in the presence of
ethylene or propylene oxide is alkylation of a component of DNA.
Not surprisingly, this mechanism results in a first-order decline in
a bacterial population exposed to the chemical with time. In this
respect chemical sterilization is identical to thermal sterilization,
and the mathematical development on sterilization processes discussed
in the previous section is applicable with appropriate values for the
parameters.

Lui et al. [15] evaluated the sterilization parameters D and z
for spores of Bacillus subtilis exposed to a mixture of 12% ethylene
oxide and 88% inert carrier at 33% relative humidity over a temperature range of 104 to 176°F. The parameter D has a value of 15 min
at 104°F and declines exponentially to 0.7 min at 176°F. The parameter z was found to be 53 F. A common value for z in moist heat
sterilization is 18 F. These data were obtained on bacterial spores
dried from sterile water onto paper and glass disks. The parameter
values obtained were not significantly different for the paper or
glass disks. These data indicate that relatively high-speed sterilization can be effected by ethylene oxide at intermediate temperatures above ambient temperature but below temperatures that would
cause serious thermal damage to solid foodstuffs.

McConnell and Collier [16] have patented a process for high-
speed sterilization of containers prior to aseptic packaging. They
propose a mixture of epoxide and water vapor at temperatures up to
210°F. These mixtures apparently become increasingly lethal to bacterial spores having high thermal resistance as epoxide concentration
and temperature increase Interestingly, pure epoxide in the absence
of water vapor is much less lethal than mixtures of 98% epoxide and
water vapor. Previous investigators [17] have placed the optimal
water vapor concentration for epoxide lethality at 30-35% relative
humidity at ambient temperatures.

C. BIOCHEMICAL CONSIDERATIONS

Besides microbial spoilage, solid foodstuffs that are maintained
in long-term storage are subject to potential spoilage by biochemical
degradation reactions. As with microbial spoilage, the role of product moisture is highly important. Microbial spoilage, while most
important in economic terms, is generally limited to products that
are in moisture equilibrium with an atmospheric water activity in excess of .70. Biochemical degradation reactions can occur at almost
any product moisture content. The type and rate of reaction is dependent on product structure, composition, moisture content, temperature, oxygen concentration, and other environmental factors.

Historically, little attention has been paid to degradation reactions of foodstuffs during storage unless they resulted in economic
loss or produced organoleptically objectionable byproducts. Now, however, more concern is evident for the decline in nutritional value
of foods during storage. Labuza [18] has provided a recent review
of nutrient losses during drying and storage of dehydrated foods.

Proper storage conditions in terms of maintenance of optimum
product moisture and temperature in addition to control of the chemical composition of the storage atmosphere can substantially influence long-term product stability.

In naturally occurring solid foodstuffs such as whole grains,
tuberous vegetables, and some fruits, product stability is favored
during storage by several natural mechanisms. Product enzymes are
physically separated from substrates that can be reactive as long as
the native cell structure is intact. Water vapor is adsorbed to surfaces where chemically active sites might be oxidized by exposure to
air. Surface area itself is limited by the intact structure of the
product.

The demands for convenience foods, and unique functional properties in food ingredients have led to many prefabricated solid food
ingredients. The prefabricated products often have an artificial
structure that may be quite porous. Examples of such food products
or ingredients are ready-to-eat cereals, textured vegetable proteins,
pre-cooked "instantized" cereals, and vegetable protein meat analogs.
Other food ingredients such as solvent extracted oil seeds, while
not prefabricated, share many of the storage characteristics of prefabricated ingredients. Many dehydrated foods also have the same
characteristics.

These solid foodstuffs may not be protected from chemical or
biochemical reactions that limit product stability by any of the
natural mechanisms mentioned previously. Any active enzymes present
after processing and fabrication are not necessarily separated from
available substrates. Unsaturated lipids and other oxidizable components may be spread over internal or external surfaces and exposed

to oxidation. The available surface area per unit weight may be
large due to the porous structure.

Development of product instability (excluding microbial growth)
in solid food products during storage is most commonly detected organoleptically by loss of flavor and color and the development of objectionable rancid odors. The chief cause of these properties is
lipid deterioration due to oxidation or hydrolysis.

Lipid oxidation can take place in foodstuffs when unsaturated
lipids are exposed to atmospheric oxygen in the presence of sufficient
temperature or the catalytic action of metals. The unsaturated lipid
molecule can react with molecular oxygen to form a hydroperoxide which
may subsequently react with more unsaturated lipid to form epoxides
and other degradation products. Peroxides decompose to offensive, volatile compounds. An excellent review [19] of the kinetics of lipid
oxidation in foods was presented by Labuza in 1971.

Numerous investigators have shown that the rate of oxidation and
the development of rancidity are accelerated in solid food products
when stored at very low moisture levels. Pigments are oxidized and
color fades. Vitamin activity can be lost. Salwin [20] has shown
that there is excellent correlation between the minimum product moisture content for product stability and the monomolecular moisture content calculated by the Brunauer, Emmett and Teller adsorption theory.
An explanation for this phenomenon is that the moisture content calculated as the monolayer is that quantity of water adsorbed directly
to all available reactive sites on all of the product surfaces which
prevents reaction of these sites with oxygen by interference.

The Brunauer, Emmett and Teller equation for equilibrium isotherms and calculation of the monomolecular moisture quantity from
an equilibrium moisture isotherm is discussed in Sec.II-A. Figure
4 locates the monomolecular moisture content of freeze-dried beef
powder, cornstarch, and corn gluten on equilibrium moisture desorption isotherms taken at 77°F. It can be noted from Fig. 4 that the
calculated monomolecular moisture contents all lie at approximately
the same location on the isotherm; near the lower limit of the linear

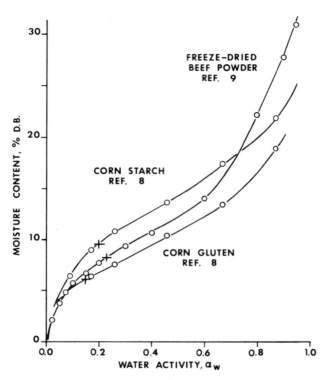

Fig. 4. Equilibrium moisture isotherms of typical food products indicating the location of the monomolecular moisture layer.

portion of the sigmoidal-shaped isotherm. Water at lower moisture contents is all adsorbed directly to the product by polar bonds.

At moisture concentrations above the limit of polar bonding in the linear portion of the isotherm, water vapor is adsorbed in multi-molecular layers. The bond strength of adsorption in each subsequent layer is less. Duckworth et al. [21] demonstrated that the mobility of water vapor and soluble components of a solid product increases with increasing equilibrium moisture content in the region of multi-molecular adsorption. Acker [22] has demonstrated that the rate of enzymatically catalyzed hydrolysis reactions increases from approx-imately zero in the monomolecular region in proportion to the amount

of adsorbed water in the multimolecular adsorption region. He con-
cludes that water serves as a medium for enzyme reactions and as a
transport vehicle for the substrates or products of the reaction.
Hydrolysis reactions are rarely if ever limited by lack of water to
directly enter the reaction. Sufficient water is available to stoi-
chiometrically enter enzymatic hydrolysis reactions even at very low
moisture levels. When the substrate is nonaqueous or can move in a
nonaqueous liquid, enzyme reactions can take place independent of the
moisture content.

From these and many other investigations it can be concluded
that the region of maximum product stability based on limited chemical activity of water vapor is in the intermediate region between
.20 and .70 on the water activity scale. Prefabricated intermediate
moisture foods for humans or pets are normally formulated to reach
equilibrium at ambient temperatures near the upper limit of this
region. By eliminating enzyme activity in the ingredients of the
product before fabrication and by addition of mycostatic preserva-
tives to suppress mold growth it is possible to push the upper limit
of stability somewhat beyond .70 water activity. This course of
action is generally dictated by increasing organoleptic appeal of
the product with increasing equilibrium water activity.

Some raw foodstuffs such as apples or potatoes for chipping
are stored in controlled atmospheric storage units where the chemical composition of the atmosphere is changed to limit the rates of
degradation reactions that depend upon oxygen. The products are
stored in crates or bulk in hermetically sealed warehouses. The
atmosphere may be allowed to change to a carbon dioxide rich, oxygen
depleted condition by natural respiration of the product or it may
be artificially changed. Artificial changes in the storage atmosphere are generally created by displacing oxygen with nitrogen or
carbon dioxide. Since the process of natural depletion of oxygen by
respiration is rather slow, artificial control of the atmosphere is
more effective but more costly.

The disadvantage of controlled-atmosphere storage is that it
cannot easily move with the product during distribution and once

the hermetic seal on the storage chamber is broken the stability-
controlling conditions are lost.

From the foregoing, it is obvious that control of product moisture and temperature represents the practical means of maximizing
storage life of most solid foodstuffs. In the following section,
the relationship of product stability and handling to the physical
design of storage facilities will be discussed.

III. BULK STORAGE AND GRAVITY FLOW

As previously mentioned vast quantities of solid foodstuffs are
stored in bulk containers in the U.S. and throughout the world.
These facilities are commonly deep bins of concrete, steel, or wood
construction. The cross section may be round or rectangular. Gravity flow is generally used to unload the bins through a hopper bottom.

It is desirable that a storage bin be designed to discharge a
particular product or group of products by gravity with even flow,
free of stoppages without leaving dead spots in the bin. It is safe
to say, however, that more bins have been designed to fit into a convenient arrangement or desirable position in the overall structure
than have ever been designed to give reliable, even gravity flow.
Attempts are usually made after construction to provide adequate
flow of the product with various types of vibrators, volumetric expansion devices, and other flow aids. A complete analysis of each
situation is required to determine if the cost of installing a properly designed bin is warranted by savings in operating costs when the
bin is in use. Bins that flow freely by gravity generally have relatively steep hopper walls, the construction of which may be more
costly than other possible designs. Certainly bins that supply
feeders which formulate final food products subject to label guarantees or which feed continuous, automatically controlled processes
should freely and evenly discharge the contents by gravity.

Operating experience has demonstrated that some bins discharge smoothly, with an even flow in which the entire mass of the load moves downward simultaneously. This is called mass flow. Mass flow provides a first-in, first-out storage regime, which is very important and desirable. Other bins discharge only from the area over the hopper opening forming one elongated vertical pipe that removes product from the top of the bin first and may periodically collapse. This is called pipe, or funnel, flow and can result in erratic discharge from the hopper. Certain products may bridge over the hopper opening, and flow will cease altogether.

Numerous investigators have considered the problem of pressures and flow of granular solids in deep storage bins, with the objective of providing mathematical expressions useful for design purposes. In 1895 Janssen [23] published a study that included a mathematical derivation for the lateral pressure of grain on the walls of a deep container. However, it was not until the 1950's that Jenike [2,3] developed and tested a theory of gravity flow of bulk solids which allows rational design of deep-bin storage facilities that provides a first-in, first-out flow regime. This theory establishes conditions for the design of hoppers and openings that given even, nonbridging flow.

Since all foodstuffs have, to some extent, a limited storage life it is highly desirable that bulk storage facilities not include dead spots that might detain product for long periods of time. Such dead spots may encourage moisture condensation, microbial growth, oxidative rancidity, or other deteriorative mechanisms that were discussed in Sec. II.

A. PRESSURE IN DEEP BINS

The pressure exerted by granular solids on the walls of deep storage bins can be very large. Because deep beds of granular solids can transfer stresses while liquids cannot, the typical expressions of density times depth for pressure at a given depth in a liquid are not accurate for granular solids. It was early observed that pres-

sure on the bottom of a bin did not increase beyond a certain value
as depth of the solids increased. Obviously the additional load is
carried by the walls. Lateral pressures on the bin walls contribute
a bending load; while vertical loads transferred to the walls by
friction of the granular solid cause column loads in the walls.

For relatively shallow bins where depth is less than two times
bin diameter lateral stress can be computed as

$$\sigma = \rho D \tan^2 (45 - \phi_i/2) \qquad\qquad (14)$$

where: ρ = product bulk density, D = depth of product below the
surface, and ϕ_i = angle of internal friction, degrees.

For deep bins the lateral pressure on the bin wall at any
depth D can be calculated from Janssen's equation

$$\sigma = \frac{R}{f_w} \left[1 - e^{-(kf_w D)/R} \right] \qquad\qquad (15)$$

where: R = hydraulic radius of the bin (area/circumference), f_w =
coefficient of friction on the wall surface, and k = $(1 - \sin \phi_i)/$
$(1 + \sin \phi_i)$. Janssen derived the equation from work on deep-bed
grain storage in 1895. It is based on considerations of static
loads. It has been stated that the pressure values computed by
Janssen's equations are greatly exceeded during dynamic situations
such as loading and unloading of a bin. However, improved equations
based on dynamic theory have not come into general use and Janssen's
equation is still widely used in the design of deep bins.

B. BASIC DESIGN DATA

As is the case with most other design data for the physical
properties of foodstuffs, tabulated data useful for the design of
deep-bin storage facilities are scarce. In general, three types of
data are required:

1. The coefficient, or angle, of friction of the product on the wall of the storage structure.

2. The effective angle of internal friction in the product mass.

3. The flow function; a plot of the unconfined yield stress versus the principal stress in the granular mass.

Both (1) and (2) are obtained from a family of yield loci, which are plots of the shear stress required to cause failure in the granular mass under compressive load. As would be expected with biological materials, all of these basic parameters are functions of moisture content, temperature, and composition of the product.

1. Angle of Wall Friction

Sliding friction of solid foodstuffs on common material of construction is documented for some products, mostly cereal grains (see references 1 and 2c). Commonly, these data are presented as the coefficient of sliding friction f_w, the tangent of the angle of friction. The static coefficient of sliding friction is that resistance which must be overcome to initiate movement of the product on the wall. This is usually a larger value than the kinetic coefficient of friction, which is the force required to sustain movement.

Figure 5 indicates the effect of moisture content on sliding friction of several cereal grains on steel and concrete surfaces. Over the range of moisture contents normally encountered in long-term storage of cereal grains, increasing the moisture content increases the friction coefficient.

Bin design must be based on the most restrictive conditions that are likely to be encountered. In general, this procedure requires that the designer select the largest angle of wall friction, ϕ_w, that is consistent with the conditions expected. The larger the angle of wall friction, the steeper the hopper walls must be to sustain mass flow of the contents. Coefficients of wall friction vary considerably with the material of the wall, however. Figure 5 indi-

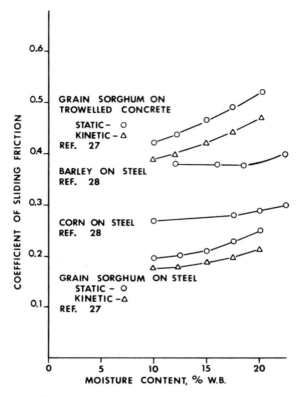

Fig. 5. Coefficients of sliding friction for selected cereal grains on various construction materials.

cates that the difference between static and kinetic coefficients of friction is less than the variation that may be expected because of changes in moisture content of the product, at least for cereal grains. If a bin is being designed for a single product it is possible, and may be desirable, to experimentally determine the coefficient of friction between the product and the wall with considerable accuracy to prevent needless overdesign of the storage facility. Detailed methods of determining friction coefficients are given in references 1 and 26.

2. Angle of Internal Friction

The angle of internal friction is the arctangent of the coefficient of internal friction which measures the resistance to slip of
a granular solid on itself. Figure 6 shows data by Lorenzen [25]
for both the angle of repose ϕ_r and the angle of internal friction
ϕ_i for wheat as a function of moisture content. The angle of repose
is often used as an estimate of the angle of internal friction, however, no fundamental, specific relationship has ever been demonstrated
that could be used to predict one angle from the other. Jenike [2,3]
has defined an effective angle of internal friction δ which is used
in the design of hoppers and openings. For real, nonideal solids

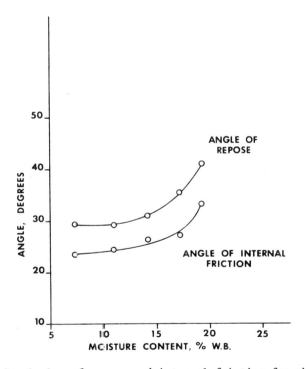

Fig. 6. Angles of repose and internal friction for wheat.

$$\delta > \phi_i \tag{16}$$

They are equal for theoretical, idealized, free-flowing solids. Determination of the effective angle of internal friction will be discussed in Sec. III-B-5.

3. The Yield Locus

Implementation of Jenike's theory for the design of hoppers and
openings is based largely on knowledge of the flow function of the
product. The flow function is generated from experimental data in
the form of a family of yield loci, which are plots of the critical
shear stress versus compressive stress. Figure 7 shows typical plots
of the yield loci. Curve A represents the yield locus of a theoretical, perfectly free-flowing, granular solid. This type of idealized
product has no cohesion, that is, the yield locus passes through the
origin, and the critical shear stress required to cause failure is
proportional to the compressive load. As a result the angle of internal friction is a constant that is not dependent on the compressive stress and is equal to the effective angle of internal friction.

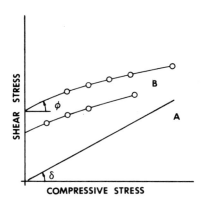

Fig. 7. Yield loci for ideal, free flowing (A) and real, cohesive (B) granular solids.

Real, granular products exhibit a family of yield loci as depicted by curves B in Fig. 7. Such products are cohesive; a finite
shear stress is required to cause failure even when they are not compressed. The yield locus of a real granular solid is dependent on
the state of compaction or consolidation existing in the mass at the
time of shear. A separate yield locus is generated for each consolidation stress. In real solids the yield locus is concave upward and
the angle of internal friction declines with increasing compressive
stress. Powders tend to have a less concave yield locus than do
coarse materials.

Jenike [2,3] designed and tested a device and procedure for determining the yield locus of granular solids. The device, shown
schematically in Fig. 8, consists of a split ring test cell 3.75 in.
in diameter, a load cell with recording equipment, and a drive device
to shear the sample at a constant strain rate.

To insure reliable accurate, numerical data for the yield locus
a detailed procedure for conducting shear tests was developed. This
procedure, adapted to solid foodstuffs, is described as follows:

(1) A uniform, representative sample of solid foodstuffs, which
has been equilibrated to the desired moisture content, is loaded into the test cell equipped with a packing mold.

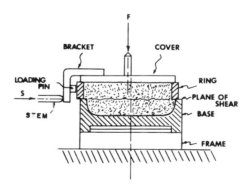

Fig. 8. The test cell.

(2) With the shearing ring offset about 0.1 in. a consolidation load L is placed vertically downward on the sample and the mold is twisted to pack and consolidate the sample. The load L creates a consolidation stress σ_c, in the sample

$$\sigma_c = \frac{L}{A} \tag{17}$$

where A is the area of the test cell.

(3) The molding cap is removed and the sample is sheared under the same consolidation stress σ_c until the shearing force S equilibrates. The equivalent shearing stress τ is computed

$$\tau = \frac{S}{A} \tag{18}$$

(4) Smaller compressive loads, L_i, are placed on the sample and shearing is repeated to determine the shear stress that will cause failure under each compressive stress.

Prior to each successive shear test the sample is compacted as in step (3) until a shear stress of about 95% of the critical stress is reached with the original consolidating load. This is to insure a uniformly packed sample in the same stress condition before each shear test.

One yield locus is generated from each series of shear tests conducted with a single consolidating load by repeating step (4) of the above procedure. The data are plotted as in Fig. 7 to give a yield locus.

The yield locus of a granular solid is affected by the time during which the sample has been under a compression load. If data are required for a product that is to be held in storage for some time it is necessary to modify the above procedure. This is done by holding the sample, under the consolidation load, in a sealed chamber at the desired temperature. The holding time must be equal to the holding time expected in the storage structure, or until an equilibrium is reached in the sample. In general, larger shear stresses are required to cause failure in products that have been held under compression for some time. With foodstuffs precautions must be taken to insure the sample does not change during the holding period due to moisture migration, microbial growth, or other chemical or physical changes.

From each yield locus are calculated the unconfined yield stress, σ_u and the major principal stress σ_1. These data are in turn used to plot the flow function of the product in question.

4. The Flow Function

The flow function of a granular solid is a plot of the unconfined yield stress versus the major principal stress. Both of the quantities can be computed from each yield locus. The unconfined yield stress can be thought of as the stress required to cause failure in an unsupported column of the solid in question.

The major principal stress σ_1 and the unconfined yield stress σ_u are computed from the yield locus by use of the Mohr semicircle technique as shown in Fig. 9. On each yield locus a Mohr semicircle is drawn through the point determined by the consolidating stress and tangent to the yield locus. The intersections of the Mohr circle with the axis of the compressive stress determine the major and minor principal stresses. A second Mohr circle is drawn through the origin and tangent to the yield locus. Intersection of this Mohr circle with the same axis determines the unconfined yield stress. Each yield locus provides one point on the flow function.

The flow function of a typical granular foodstuff is shown in Fig. 10. The flow function for the same material equilbrated under compressive load for time t is also shown.

5. Effective Angle of Internal Friction

The effective angle of internal friction δ is defined by the abscissa and a line from the origin tangent to the Mohr circle, which defines the principal stresses. See Fig. 9.

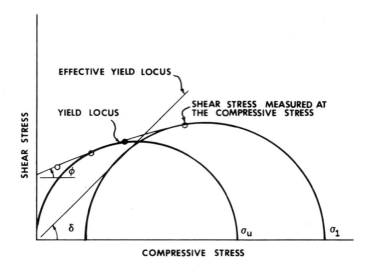

Fig. 9.   Analysis of a yield locus.

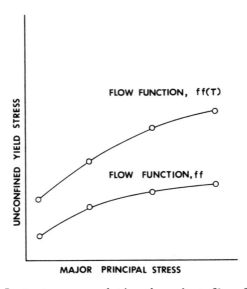

Fig. 10.   Instantaneous and time-dependent flow functions.

C. DESIGN OF HOPPERS, OPENINGS, AND CHUTES

Section III continues with methods of design for hoppers and openings based on the work of Jenike [2,3]. The theory developed by Jenike is based on the principles of plasticity and only the equilibrium situation is considered. Forces due to momentum transfer are neglected as being insignificant compared to body forces. The practical effect of this limitation is that the method can only be applied when the velocity of material is low. It does not apply to bins unloading at a free-flow rate into unrestricted ducts or tanks. For most applications this is not a significant limitation since bins usually feed some sort of flow controlling or metering device that restricts exit velocity far below the free-flow rate.

According to Jenike's method of design, the flow properties of a particular solid, as determined in the previous section, are related to the shape and material properties of the hopper and opening to determine conditions that assure flow. Gravity flow exists as long as no stable structure of the granular solid can form over or above the opening. If flow is stopped, one of two types of structure usually develops: either a dome over the opening or a vertical, empty pipe extending from the opening to the surface with diameter approximately equal to the major dimension of the opening.

Jenike [2] analyzed the incipient failure condition of both stable structures in a granular solid mass. Let σ_s be the surface stress applied to a stable structure in a granular mass. Recall that σ_u is the unconfined yield strength a solid develops at an exposed surface and that

$$\sigma_u = f_u(\sigma_1) \tag{19}$$

is the flow function of a granular solid. For a structure to become unstable the following inequality must be maintained:

$$\sigma_s > \sigma_u \tag{20}$$

That is, to cause failure the surface stress must exceed the unconfined yield stress. By dividing both sides of Eq. (20) we have

$$\frac{\sigma_s}{\sigma_1} > \frac{\sigma_u}{\sigma_1} \tag{21}$$

Jenike defined the flow factor ff to be the ratio,

$$ff = \frac{\sigma_1}{\sigma_s} \tag{22}$$

Thus, to insure instability of any structure in the granular mass,
the inequality

$$ff > \frac{\sigma_1}{\sigma_u} = \frac{\sigma_1}{f_u(\sigma_1)} \tag{23}$$

must be maintained.

By theoretical analysis of the incipient failure condition the
flow factor ff has been determined as a function of:

(1) the effective angle of internal friction δ,

(2) the angle of wall friction ϕ_w, and

(3) the slope of the hopper wall θ (i.e., the angle between the
hopper wall and the vertical.

Jenike [3] presented numerous plots of the critical flow factor

$$ff = ff(\delta, \phi_w, \theta) \tag{24}$$

A typical example for a symmetical conical hopper is shown in Fig. 11.
By knowing the effective angle of friction, the angle of wall friction, and the slope of the desired hopper the flow factor can be determined. Use of the flow factor in design of hoppers is discussed
in the next section.

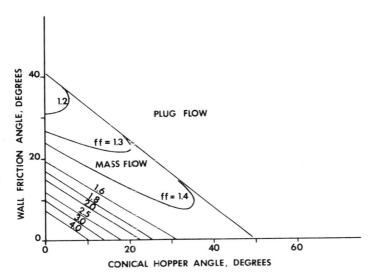

Fig. 11. Critical flow factors of conical hoppers for materials with effective angle of internal friction of 50°.

A guide to the general flow characteristics of a granular solid can be gained from the following inequalities.

$$\frac{\sigma_1}{f_u(\sigma_1)} > 10 \qquad \text{free flowing}$$

$$10 > \frac{\sigma_1}{f_u(\sigma_1)} > 4 \qquad \text{easy flowing}$$

$$4 > \frac{\sigma_1}{f_u(\sigma_1)} > 2 \qquad \text{cohesive}$$

$$2 > \frac{\sigma_1}{f_u(\sigma_1)} \qquad \text{nonflowing}$$

1. Design of Mass-Flow Hoppers

Two types of flow exist in gravity unloading of bins, mass flow and funnel, or plug, flow. Jenike [3] discusses the rationale for the use and design of plug-flow bins but this type of flow is undesirable for foodstuffs and will not be considered here. In fact, Jenike [3] makes the statement, "...the main reasons why plug-flow bins are so commonly used today are: the general lack of knowledge of the existence and advantages of mass-flow bins, and the notion that plug-flow bins are cheaper per unit volume. That economy is more apparent than real: it often becomes quite expensive in terms of excessive startup time, disrupted production schedules, additional shifts, lower product quality, unrealized storage, redesigns, modifications, flow promoting devices and time spent by operating and supervisory personnel in keeping the solids flowing."

Mass-flow hoppers can have various shapes but are generally characterized by having relatively steep hopper walls and corner valleys that flow out at the opening. Mass flow is characterized by uniform discharge rate, constant bulk density, minimum attrition to the product, minimum wear on the walls, no dead regions, first-in first-out flow regime, and no segregation during storage.

Mass flow theoretically exists for all combinations of hopper wall angle θ and angle of wall friction ϕ_w which lie within the envelope of flow factors shown in plots like Fig. 11. For practical, conservative design it is desirable to stay 3 to 5° within the envelope on the hopper angle scale to allow for changes in the wall friction angle.

For each angle of wall friction there exists a minimum, optimum flow factor. The selected flow factor can be plotted over the flow function of the product in question to determine the dimensions of the hopper outlet. See reference 3 for a complete selection of flow factor charts for conical and plane flow channels.

2. Design of Openings for Mass Flow

In mass flow only the minimum opening dimension B, which prevents doming, is critical. This dimension is selected by plotting the flow factor ff of a selected hopper over the flow function of the material in question. If the flow factor ff lies below the flow function, the minimum dimension is not determined on the basis of doming, but must be selected for desired rate of flow and other factors.

If the flow factor intersects the flow function the critical unconfined yield strength σ_{cu} is defined by the intersection and the minimum dimension of the opening is determined by

$$B = \frac{H\sigma_{cu}}{\rho} \qquad\qquad (25)$$

where H is determined from Fig. 12 as a function of the hopper angle.

If the flow factor lies entirely above the flow function gravity unloading is impossible and a new design for the hopper must be selected.

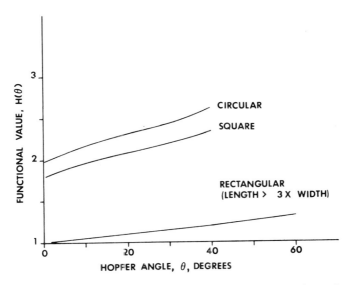

Fig. 12. The function H for design of hopper openings for mass flow.

3.   Gravity Flow in Discharge Chutes

The equilibrium analysis used for hopper and opening design does
not result in an evaluation of the velocity of the material during
discharge. Thus, it does not provide any information on the required
opening size needed to deliver a given rate of discharge.

Roberts and Arnold [26] have analyzed gravity flow of granular
solids in discharge chutes. They determined two modes of flow in
closed ducts: fast flow and slow flow. Fast flow exists when the
flow mass does not contact the top of the duct. In this case flow
rate is independent of the radius of curvature or other dimensions
of the duct and is controlled only by the size of the hopper opening.
If the duct geometry is such that a momentary obstruction can cause
the granular solid to contact the top of the duct the flow rate immediately drops and slow flow exists. In slow flow the rate is controlled by the geometry of the duct and the friction factor of the
product sliding in the duct.

Stewart [27] determined from analytical and experimental investigations that parabolic chutes give preferred performance.

In parabolic chutes Eq. (26) applies,

$$y = cx^2 \qquad (26)$$

where x = distance below the outlet, y = lateral distance, and c =
constant. The constant c must be such that the angle of slope at the
end of the chute is less than the critical angle θ_c

$$\theta_c = 90 - \tan^{-1} f_w\left(1 + \frac{kh}{b}\right) \qquad (27)$$

where f_w = friction coefficient on the duct wall, h = height of duct,
and b = width of duct.

REFERENCES

1. N. N. Mohsenin, Physical Properties of Plant and Animal Materials, Vol. 1, Gordon, New York, 1970.

2. A. W. Jenike, Bulletin No. 108, Utah Engineering Experiment Station, University of Utah, Salt Lake City, Utah, 1961.

3. A. W. Jenike, Bulletin No. 123, Utah Engineering Experiment Station, University of Utah, Salt Lake City, Utah, 1964.

4. G. D. Saravacos and R. M. Stinchfield, J. Food Sci., 30, 779 (1965).

5. B. Makower and G. L. Dehority, Ind. Eng. Chem., 35, 193 (1943).

6. I. Langmuir, J. Am. Chem. Soc., 40, 1361 (1918).

7. S. Brunauer, P. H. Emmett, and E. Teller, J. Am. Chem. Soc., 60, 309 (1938).

8. D. S. Chung and H. B. Pfost, Trans. ASAE, 10, 522 (1967).

9. P. O. Ngoddy and F. W. Bakker-Arkema, Trans. ASAE, 13, 612 (1970).

10. D. A. A. Mossel and M. Ingram, J. Appl. Bacteriology, 18, 232 (1955).

11. C. R. Stumbo, Thermobacteriology in Food Processing, Academic Press, New York, 1965.

12. L. Sair and H. J. Pappas (to Griffith Labs.), U. S. Patent 3,206,275 (1965).

13. H. Bakerman, M. Romine, J. A. Schricker, S. M. Takahashi, and O. Mickelsen, Agri. Food Chem., 4, 631 (1970).

14. G. T. Michael and C. R. Stumbo, J. Food Sci., 35, 631 (1970).

15. T. S. Lui, G. L. Howard and C. R. Stumbo, Food Tech., 12, 86 (1968).

16. J. E. W. McConnell and C. P. Collier, Food Eng., 34, 96 (1962).

17. G. L. Gilbert, V. M. Gambill, D. R. Spiner, R. K. Hoffman, and C. R. Phillips, Am. Soc. for Microbiology, 12, 496 (1964).

18. T. P. Labuza, Critical Rev. Food Tech., 2, 355 (1971).

19. T. P. Labuza, Critical Rev. Food Tech., 3, 217 (1972).

20. H. Salwin, Food Tech., 13, 594 (1959).

21. R. B. Duckworth and G. M. Smith, Recent Advances in Food Science,
     3, 230 (1963).

22. L. W. Acker, Food Tech., 23, 1257 (1969).

23. H. A. Janssen, Zeitschrift des VDI, 39, 1045 (1895).

24. Agricultural Engineering Yearbook, ASAE, St. Joseph, Mich.

25. R. T. Lorenzen, Masters thesis, Univ. of California, Davis,
     Calif., 1957.

26. A. W. Roberts and P. C. Arnold, Trans. ASAE, 14, 304 (1971).

27. B. R. Stewart, Q. A. Hossain, and O. R. Kunze, Trans. ASAE, 12,
     415 (1969).

28. W. G. Bickert and F. H. Buelow, Trans. ASAE, 9, 129 (1966).

Numbers in parentheses are reference numbers and indicate that an author's work is referred to although his name is not cited in the text. Underlined numbers give the page on which the complete reference is listed.